James Peattie

Steam Boilers

Their Management and Working on Land and Sea

James Peattie

Steam Boilers
Their Management and Working on Land and Sea

ISBN/EAN: 9783743465138

Manufactured in Europe, USA, Canada, Australia, Japa

Cover: Foto ©berggeist007 / pixelio.de

Manufactured and distributed by brebook publishing software (www.brebook.com)

James Peattie

Steam Boilers

STEAM BOILERS:

THEIR MANAGEMENT AND WORKING ON LAND AND SEA.

STEAM BOILERS:

THEIR MANAGEMENT AND WORKING ON LAND AND SEA.

BY

JAMES PEATTIE.

E. & F. N. SPON, 125, STRAND, LONDON.
NEW YORK: 12, CORTLANDT STREET.
1888.

PREFACE.

As nearly all the writers on this subject have been non-practical, therefore, notwithstanding their scientific attainments, reasonings, and data being ever so scientifically correct as a whole, their writings are void of many important details and important practical facts; and being couched in abstruse and technical language, are unintelligible to the mass of persons in charge of steam boilers, and uninteresting to others, who fail to find in them the particular counsel they desire.

Under such circumstances it is obvious that much conflict of opinion and conflict of practice will exist amongst those in charge of steam boilers, and even among practical men, regarding the subject. Any effort, therefore, to treat the subject in a thoroughly simple yet practical manner, in the language of the engine-room, the workshop and the factory, will be welcomed by those having a *bonâ-fide* interest in the management and running of steam boilers.

My qualifications for writing on this subject may be inferred from my having had many years' practical experience in the management and running of boilers, as engineer in charge of works and factories on shore, and chief engineer of well-known vessels at sea. Having been associated with the steam boiler over a period of thirty years, I have endeavoured to write up unwritten experience in my own style, and not in the stereotyped style usual with such subjects, and I trust the contents may be found to some extent profitable.

JAMES PEATTIE.

32, SCOTIA STREET, GLASGOW.

CONTENTS.

	PAGE
INTRODUCTORY	1
WATER	4
WATER AND SOLIDS	8
COAL	8
COAL GASES	12
COMBUSTION	15
GENERATION OF STEAM	25
INCRUSTATION OR SCALE	27
WATER ANALYSIS	33, 40
ELEMENTS OF NATURE	34
COMBINING QUANTITIES	37
SCALE ANALYSIS	38
DENSITY—COMBINATION OF SOLIDS	41
SPECIFIC GRAVITY AND DENSITY	43, 50
DENSITY AND BOILING POINTS	44
SCALE FORMATION	45
BLOWING PRACTICES	46, 95
BLOWING OUT WATER FOR FRESHENING	48
DEPOSITION OF SOLIDS	51, 54
SALINOMETER	52, 59
CHARACTER OF SCALE	52
SALT DEPOSIT	56, 188
SCALE-DEPOSITING	57
FURNACE COLLAPSE	58
VIRTUES OF GOOD CONSTRUCTION	59
DENSITY	61
ARTIFICIAL SCALING	62
FLUXING SURFACES	63
SALT-SEASONED	64
PRIMING	65
CIRCULATION	73
PITTING	74
STIFF FOR STEAM	78
SOOT	82
SOOT AND SCALE EFFECTS	83
FEED	91
BLOW-OFF COCKS	94
BLOWING OUT WATER	95
CHANGING WATER	97
WASHING OUT	99
BLOWING	100
RETENTION OF WATER	101
ACIDITY	102
SODA	104
ZINC	112
SCALE PREVENTION	117
BOILER MEDICINES	118
WATER EXPANSION	120
EXPANSION OF BOILERS	123
LATENT HEAT	124
SUPERHEATING	138
BOILER FILLING	139
THE GAUGE-GLASS	140
BOILERS LIFTING THEIR WATER	142
FURNACES	142
FIRING	143
SMOKE	147
DRAUGHT	148

Contents.

	PAGE
VENTILATION	149
DOWN-DRAUGHT	150
CHIMNEY-TOP ABLAZE	152
FIRE-BARS DROPPING	153
DAMPERS	154
DAMPER GEAR	155
EASING FIRES	156
SUDDEN STOPPAGE	156
KEEPING STEAM HANDY	157
ALARUMS	157
SAFETY-VALVES	158
BOILER GAINING WATER	159
LEAKY BOILER	159
HYDROKINATOR	160
VACUUM IN BOILER	161
CORROSION	162
RUSTY BOILERS	164
CHIPPING HEATING SURFACES	164
BANKING FIRES OR DAMPING	167
HOLING FIRES	167
TUBES BURSTING AT SEA	168
LEAKY TUBES	169
TUBE STOPPING	170
CONCENTRATED FLAME	174
BRIDGES	175
GALLOWAY TUBES	176
CONCENTRATION OF HEAT	178
CONCENTRATION OF FLAME	180
UPRIGHT BOILERS	181
LAND BOILERS—	
EGG-END	182
CORNISH	184
BOILER REPAIRS	185
TEAR AND WEAR	193
GROOVING	194
PRESSURE GAUGE	195
VACUUM GAUGE	197
PRESSURE TESTS	198
EXPLOSIONS	199
JOINTS AND CONNECTIONS	212
CYLINDER LUBRICANTS	214
PRESSURES	215

STEAM BOILERS:

THEIR MANAGEMENT AND WORKING ON LAND AND SEA.

INTRODUCTORY.

1. It is not the object of this work to trace the history of steam or the steam boiler from times remote to the present, or to extol the virtues of any particular kind or conformation of boiler, except in so far as is compatible with the principles involved, and absolutely necessary to the elucidation of the subject generally, and the economy and safety of the present types of boiler in particular. The history, proportions, and development of boilers have from time to time been well discussed by scientific as well as practical writers. I propose here to deal with the steam boiler as a *production* only.

2. As a matter of course hard facts relating to economy, convenience, and utility, with no doubt a little fancy, have conspired to make the present types of boilers fixtures, and to quarrel with these productions, or argue the point regarding their respective merits, is not my intention, and outside of my province.

3. The marine boilers of the present type, with

surface-condensing engines, necessarily require more to be said of them, on account of their intricate conformation, their long periods of isolation, and the great issues hanging thereon. A treatise on these will cover most of the important features of the others, and where it does not, the differences will be treated, as well as the salient points deduced from experience of each.

4. Boilers may be said to be of three kinds, viz. externally heated, internally heated, and both. Of the first we have the egg-end boiler, waggon boiler, spherical boiler, &c. Of the second the upright boiler, the circular and multitubular, as also numerous fancifully heated boilers. Of the compromise we have the dry-bottomed multitubular, Cornish or flued land boiler, locomotive or fire-box and tubed boiler. For general purposes on shore the upright form is the most common, owing to its convenience and portability. The term "vertical boiler" is often used for this class. "Donkey boiler" is quite a favourite term for all small boilers not set up in brickwork; and indeed the term is in many cases very appropriate, as in a comparison between the proverbial long-suffering, uncomplaining, and often sorely abused and neglected "cuddy," and the equally maltreated and neglected "donkey boiler," the term as applied to the boiler itself is most realistic. Similarly mistimed for fuel, food, and water, and with a similarly promiscuous class of attendants or masters, in the matter of comparison they have much in common. The one will suffer

on, until death release him; the other may withstand similar treatment under circumstances even more varied, but will certainly not yield its life, as did its more ancient relative, without resentment, or without placing its untimely end on record. Although the death of a donkey may be a rare event and have few witnesses, the bursting of a donkey boiler is by no means so rare or so easily got over. The donkey boiler will not always be neglected with impunity. No matter how menial the employment, or obscure the circumstances of the owner, or situation of the boiler itself, neglect will sooner or later speak for itself, when the inevitable explosion occurs. The troubles and necessities of the nondescript itinerant will be as little available against the result as the excuse of the wealthy manufacturer.

5. The practice of putting in charge of steam boilers boys and other persons who are as ignorant of the duties as of the risk incurred, is just as reprehensible as placing in charge of a spirited horse a person who never saw one before. It is not necessary that a boiler attendant be an adept in science, or that he be an engineer, or even a boilermaker, as neither the one nor the other is at all likely to have received during his probation any special training on the subject more than may have come in the way of any craftsman, whether mechanic or not. The man to select for a boiler attendant or engineman must be a sensible, steady, methodical, and inquisitive man, who can weigh the philosophy of common things as he goes along.

A man who has a respectful fear of a gun, loaded or unloaded, and has also a shrewd respect for another man a foot over him in stature and strong in proportion, this is the man to train for the post. But the question is, how is he to be trained? The school no doubt is a large one, but there is no teacher.

The man will have to take his cue from those around him, many of them being perhaps less intelligent than himself, but his instructors all the same. Very soon he will discover for himself that he has dropped into the same school as did the old man and his son who took their ass to market and lost it by the way, through acting upon undiscriminating advice.

6. Although there is nothing very profound or awfully scientific in the management of boilers, there is, however, no royal road to it. The circumstances generally, as well as the disposition of the boiler itself, must be studied just as any other subject in nature, even as the human subject, or nature itself.

The experiences and teachings of our mothers or grandmothers regarding the tea-kettle, the smoke cowl, the stirring of the pan, &c., are each powerful instructors on the subject of steam boilers, not only as analogous cases, but as absolutely cases in point applicable in detail.

WATER.

8. All waters used for boiler purposes on land or sea are more or less impure. Whether the hard spring water impregnated with mineral compounds,

the soft water from the bog, the filtered water of the city, or that of the ocean, all are practically impure.

The last is preferable to doubtful fresh waters for boilers. The solids held in solution by sea-water are remarkably enough similar to those of fresh, excepting salt, which is the chief index of density.

Our object is to get the water as pure as possible, as all the ingredients found in water are individually injurious in one shape or other to the boiler, and when in combination are in many cases very destructive. If we cannot have the water chemically pure, let us have it clean, that is, never admit water into a boiler that is muddy, or otherwise unclean to the eye. But whatever be the source of our supply, let us know its properties beforehand. The indiscriminate use of water for a boiler (allowing it in time to speak for itself) is very reprehensible. (See §§ 77–8.)

9. Rain water, as gathered in country districts, is the most pure, but if gathered within the boundaries of a large city is often very objectionable, owing to carbon, soot, and other impurities being gathered with it. Generally speaking, what is termed "soft water" is preferable to that which is termed "hard water," which we can tell by tasting, or by washing our hands—if the soap do not form lather but produces a sediment or curd the water contains mineral matter, such as lime and other salts. Any one having tasted mineral waters will be able readily to form an idea of its unfitness for boiler purposes. We must bear in mind that all mineral or vegetable matter that is contained in the water is deposited in the boiler; we

cannot extract the solid matter from the water, but we can evaporate the water, thus leaving the residuum. Take a tumbler of hard water and boil it until all the water be evaporated, the residuum or "grounds" remaining at the bottom will represent the minerals or solid matter contained in the water, in kind and quantity. Of course a test would be more satisfactory with a larger vessel. Now, supposing we evaporated sixty tumblers of clear water and that the residue, as collected, just fills one tumbler, then we say that the amount of solids held in solution by the water is as one in sixty, or one-sixtieth, and in figures as $\frac{1}{60}$th.

10. Now no better test comparison in a simple way could be made. As the water is put into the boiler and boiled off, leaving the solids in the boiler, just as left in the tumbler, the process in the tumbler is just a reproduction of the process in the boiler, which we will just repeat to find what is contained in *drainage water*, which includes bogs, ditches, ponds, "burns," creeks, lochs, and rivers. It will, however, be obvious that we must test the water when in its normal condition, as doing so after heavy rains and when the district is in flood is out of the question. Water so heavily charged with mud and earthy matter should not be used for boiler purposes without admitting it first into a tank, to allow the mud to precipitate or settle to the bottom; the clear water should then be drawn off from the top. Any one may judge of its fitness by the eye alone, which decision may often be overruled by the necessity of

the case. It is, however, easily borne in mind that boiler water cannot be got too clean. Now, assuming our test to be under fair conditions, the result compared with spring water will be essentially different; the sides of the tumbler or vessel will, as the water is evaporated, be coated with a *scum*, thin and very irregular, owing to the motion of the water in boiling: the surface will also have this in patches. We will also have a mineral deposit similarly as in spring water, but less in quantity. Now, if one in sixty is the proportion of solid matter, as quoted, in the water (our boiler being now filled with it, and previously clean), there is not much to complain of.

11. Our boiler is filled with sixty bucketfuls; that is, 59 buckets of water and one bucket of solids, and suppose we use six buckets of feed-water every hour, in ten hours all our first charge of water is gone, and another charge admitted, which means another bucketful of solid matter in the boiler, and this rate would fill the boiler in sixty days. But of course long ere this, steaming would be impossible; therefore, the amount of *feed-water* we use in a given time determines exactly the amount of solids taken into the boiler (see §§ 77–78), which have to be got rid of by judicious scumming (see § 161), blowing (§§ 161–167), and cleaning (see § 173); as all extraneous matter whatever admitted into the boiler is injurious; some indeed being very destructive substances, demanding the constant care and attention of the attendant, or engineer in charge. It is good practice with land boilers to have a large tank,

such as an old boiler, to store rain-water, in order, if the ordinary water is productive of scale, to dilute it therewith in feeding; failing this, a large cask, which is kept full by a self-acting ball-cock. The sediment thus falls to the bottom, the water for the boiler being drawn from the top.

WATER AND SOLIDS.

12. We must not forget that the nature of the solids is an important consideration. For instance, one kind of water may be very heavily charged with solids, and workable for steam, while another water not so heavily charged may be unfit for boiler purposes, owing to the presence in the latter of substances that the boiler cannot *digest*, such as sulphate of lime, or, substances present in such proportions as combine chemically (inside the boiler) while working, to produce scale (see §§ 40, 41).

COAL.

13. Coal may be divided into two kinds; the bituminous, and non-bituminous. The first kind includes cannel or gas coal, Scotch main and ell coal, and Newcastle, or North of England coal. We then come to the non-bituminous, as Lancashire, Derbyshire, and South Wales anthracite, this being the least bituminous of all. This cannel coal is the greatest flaming coal, being rich in hydrocarbons (such as tar, pitch, oils, and other bituminous substances), and is hard, and stands exposure best. This cannel coal may be put as the original coal

deposit pure and simple, all other coal having been partially burned in the bed, the bituminous products thereby having been more directly turned into carbon " pure," or otherwise set free.

Scotch coal comes next, being the least burned, and stands exposure well. Then Lancashire; then Newcastle, Derbyshire, Welsh, and anthracite or South Wales coal, which has been much burned or "coked" in the bed. The difference between this and coke itself is that of the one being burnt under cover naturally, and the other artificially, the bituminous or smoking and flaming element being gone in both cases, which accounts for coke and Welsh coal being alike smokeless. This coal, although rich in carbon, is poor in hydrogen, which as an element in combustion, under certain conditions, exceeds carbon itself bulk for bulk. It is therefore benefited by being slaked with water before being laid on the fire.

14. To work coals economically by the addition of hydrogen involves more skill than that of throwing bucketfuls of water over them before entering the furnace. Hydrogen being the chief element of water, as well as that of soot and smoke, is also a prime mover in combustion, and takes the initiative in combining with air. This property of taking the initiative is an all important one, although not well understood. Suppose we have carbon and hydrogen in equal parts, and air present in sufficient quantity to combine with either, but not both. The hydrogen will at once seize its combining quantity of air,

leaving the less volatile carbon to perish. Oxygen will also take the initiative in uniting with carbon, in combustion in which it figures (see § 46), similarly as the developing medium of the carbon or burnable part. Welsh coal being heavy takes less stowage, requires little skill in stoking, and is, all things considered, our best all-round furnace coal. Although its intense local heat in the furnace, with no flame, often giving out 70 per cent. of its heat there alone, is often apt to speak for itself (see § 92), the great object of getting rid of the insulation or smoke interposition is most obtained with Welsh coal. Scotch, although having more hydrogen and also more oxygen to liberate the hydrogen from the fuel, which prevents caking, although demanding more air for combustion, cannot beat Welsh coal bulk for bulk as a smokeless steam coal, and the latter is improved on being wetted with fresh water up to a certain point, which if exceeded, would only result in producing aqueous vapour unmated by its companion products, the carbon would be condensed into soot, and altogether it would be the reverse of economical.

15. Coal-caking shows that the hydrocarbons (or hydrogen in combination with carbon) has not been liberated beyond the surface; the air admitted being under *par* for flame. The fuel is partially coked. When it is undesirable for the coal to get together in this way reduce the thickness of fire; the deeper the fire is kept the more will it cake.

16. Coal is used most economically in large lumps fresh from the seam; it evaporates on exposure to the

air; as the temperature increases it cracks and opens up in the way of its cleavage; then across, each section again dividing and subdividing with the addition of heat, the structural integrity of the lump being maintained throughout. Coal exposed to sun or weather or stowed in a warm place may lose from 5 to 25 per cent. of its original value as fuel, and like other merchandise it suffers with vicissitudes of transport, all the while breaking up and crumbling into "*slack*." (See § 18.)

17. It is good practice to adapt the furnace to each description of coal, thus: for long-flaming bituminous, shorten the fire-grate with a row of bricks in front of bridge over on the bar ends. For the short-flaming, lengthen the fire surface to the extent of brick on flat of bridge only. The character of different kinds may be fairly well judged by watching the combustion in an open fire.

18. The manufacture of "screenings" and other slack into patent fuel has one good effect, that of keeping that stuff out of the market. Of course slack and other refuse containing all the sand and foreign matter in the seams is not good for burning, neither is the outer crust of the lump that suffers from exposure. The manufacture of patent fuel may well be encouraged as that rubbish has hitherto been a scourge in a vessel's bunkers. Slack coal is best utilised in land boilers by wetting to induce it to cake, then lifting the fire up to break it through with the slice.

Coal Gases.

19 A. (See §§ 20–22.)

Fig. 1.

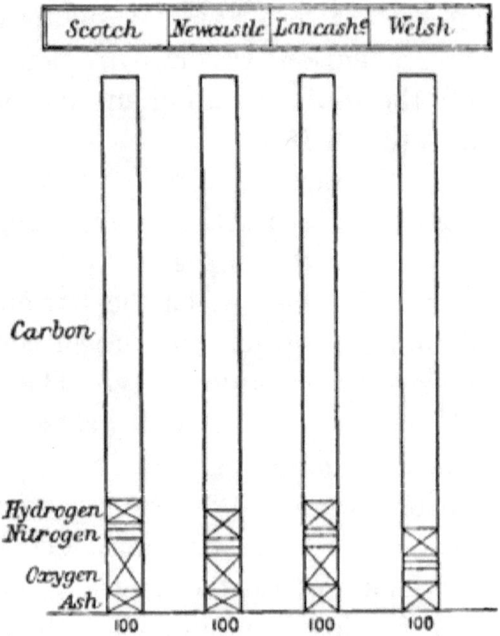
Scotch
Newcastle
Lancashe
Welsh
Carbon
Hydrogen
Nitrogen
Oxygen
Ash
100
100
100
100

B.

	or	or	or	or
Carbon	78½	81	80	85
Hydrogen ..	5½	5½	5	5
Nitrogen ..	1	1½	1	1
Sulphur ..	1	1½	2	1
Oxygen	9½	6½	7	3
Ash	4½	4	5	5

Temperature of 32° F. may be put down as most suitable in the production of steam. Every one knows that a fire burns best in cold weather, owing to the greater density and purity of the air. The fuel also is less deteriorated. The expansion of air by heat is well shown with a bladder. If this is filled by the breath in a room where there is no fire and taken to a room where there is one, the bladder will burst. If taken outside it would have been slack. Comparing air at 32° F. to unadulterated milk, and air at say 212° F. to a compound of milk and water in equal proportions, we have a fair comparison of quantity and quality, increase of the one and decrease of the other.

Fig. 2.

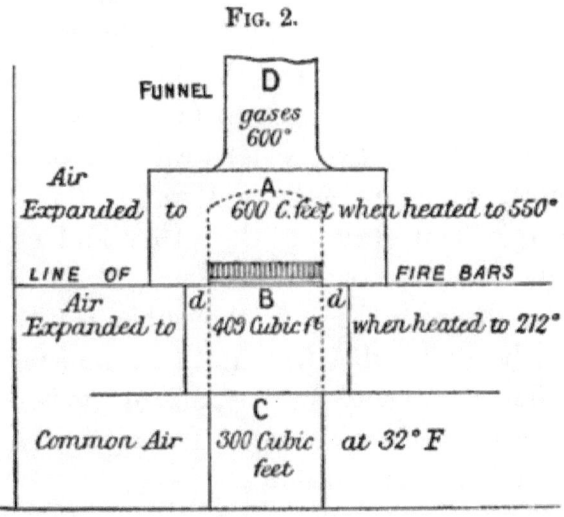

20. The sketch, Fig. 2, shows the part taken by common air in combustion. C shows the air at 32° on entering the ash-pit. B shows owing to the heat of ash-pit how the expanded air is throttled and excluded as at d, d, proving the utility of cool ash-pits. A shows the same air when heated to 550° as equal to chimney draughts with the volume increased to

double. This air is on the point of joining the coal gases in combustion. Now, we require the air through our ash-pits as cold and dense as possible, and in proportion as this air is increased in temperature so it is in volume or bulk; therefore, if the outside air be 32° and we, owing to the hot ashes in the ash-pits, allow the air to be heated to 212°, we will, as per sketch, be only receiving three-quarters of our proper quantity of fresh air. Because, as the outside air represents 100 per cent. and we dilute this to 73 per cent. of its original density, and only being able to receive each bulk for bulk, we are just parallel with the milk case: the density at first being 100 per cent., was diluted to 50 per cent. by one quart being made two, and as we can only admit one quart, which is milk and water, therefore we receive one pint, leaving the other and of course losing it.

As per diagram, Fig. 2, 300 cubic feet of air at 32° will be 409 cubic feet at 212°, and 604 cubic feet at 550°. Twenty-four pounds of air is required for each pound of coal burnt. The air in the funnel should be from 522° to 650°.

21. At 50° in the ash-pits the air in the funnel D should be 600°. The chimney temperature must exceed that of the steam in order to produce draught, which is represented by the excess. Heat releases the captive metals from the ore just as the sun develops the rose-bud.

Loss by radiation 5 to 10 per cent.

Loss by hot ashes and clinker 15 per cent.

One pound of coal yields 800 units of heat.

Combustion.

COMBUSTION.

22. There are several kinds of combustion, but we need only speak of what concerns the steam boiler.

Fig. 3 shows the relative proportions of air and

FIG. 3.

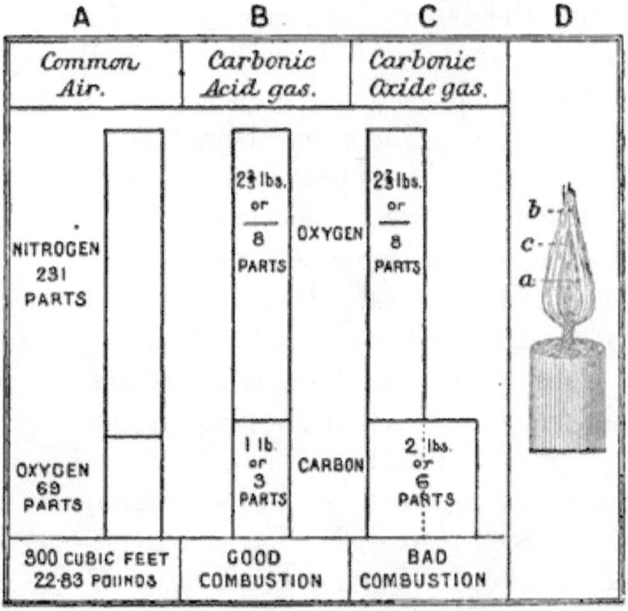

A	B	C	D
Common Air.	Carbonic Acid gas.	Carbonic Oxide gas.	
NITROGEN 231 PARTS	$2\frac{2}{3}$ lbs. or $\frac{8}{8}$ PARTS OXYGEN	$2\frac{2}{3}$ lbs. or $\frac{8}{8}$ PARTS	b — c — a
OXYGEN 69 PARTS	1 lb. or 3 PARTS CARBON	2 lbs. or 6 PARTS	
300 CUBIC FEET 22·83 POUNDS	GOOD COMBUSTION	BAD COMBUSTION	

coal gases in combustion. A shows the composition of common air. B and C show the two kinds of combustion common to furnaces. With the proportion of oxygen and carbon as at B, and on the impact of flame, the combustion is perfect, and also the most economical, and so long as the proportions here put down are maintained the smoke will be well consumed. The heat from the incandescent fuel sets free the carbon of the coal, which mates itself immediately with its proper quantum of oxygen from the air as naturally as a hungry animal seizes its food. 1 lb. of carbon to $2\frac{2}{3}$ lb. oxygen in presence of heat produces carbonic acid gas simultaneously with flame and light, whereby the products of the coal are consumed. This is the perfect form of combustion.

23. If the quantity of fresh air be reduced, or what has the same effect, the carbon increased, to the proportion of C or 2 lb. carbon to $2\frac{2}{3}$ lb. oxygen, then there is no conjugal impact of the gases—no flame, no light. The unburnt products of the coal passing up the chimney are called carbonic oxide gas. This is imperfect combustion.

A. To witness a large building on fire will give a good idea of the process. Dense columns of smoke at one point show that the flame-producing portions are disengaged faster than there is air at hand to make the combination. All at once bright flames springing up out of the darkness show that the liberation of the carbon has been checked, or that the volume of air has been increased. The smoke is being consumed, thus rendering its quota of heat. Those long, lithe, slippery flames spring, tiger fashion, from point to point, making a footing good wherever they can find their sustenance, viz. the escaping gases and air sufficient for the smoke, often existing at great distance from their proper base of operation—the incandescent fuel. This flame then is the embodiment of B, or perfect combustion.

B. The combustion in a burning building is fitful as that of the furnace. We can have an example of steady combustion from a common candle flame,

Fig. 3, D. The black cone *a*, over the wick, is hollow and contains the unburnt gas from the wick. At *c* the flame becomes luminous, but the combustion is still incomplete and illustrates carbonic oxide combustion, C. *b* is the outer zone or mantle where the combustion is complete. At this point more air has got into the mixture, producing carbonic acid combustion, B. The difference in economy of the two kinds of combustion averages 25 per cent. or one-fourth of the whole consumption. In tubed boilers this difference is greatest, from the effects of extra soot deposit consequent on the imperfect combustion.

24. The heat developed or other useful effect derived from the latent power of coal by combustion B amounts to nearly $8800°$ of heat or units of work clear of all deductions. $44,000°$ is the perfection of economy. Making allowance for the 25 per cent. loss by incomplete combustion there is still a great disparity.

A glance over the chimneys and stalks of a large city will demonstrate the two kinds of combustion B and C. The dense black smoke indicates a new laid on or green fire; change of this phenomenon shows that the combustion has changed from C to B. Great waste often occurs from the escape of coal and air gases without combustion at all, notwithstanding the proper combination. For instance, gas will escape from a burner unburnt because the flame may have been blown out, or the gas not have been ignited, or by a fire that is too dull to start a flame, which is the seal of the chemical union.

25. A smith when blowing up a large green fire, termed "charring," will in due time pierce it to the heart with the poker, making a hole. The dense smoke or gases issuing from the warm coal now get ignited, and combustion with flame and heat is established.

A candle or gas-jet cannot be lighted with an iron rod that is only red-hot; it must be nearly of a white heat. This shows that heat and flame are separate conditions, that flame is itself an element in combustion, and that coal and air gases, in proportions of either B or C, will escape without combustion in the absence of flame. With good combustion, as B, the escaping gases carry with them so much heat as is necessary to produce draught; but bad combustion not only carries the heat away, but also carries away bodily the material the heat is made of, in the form of dense black smoke. There is no milk and water or half-and-half combustion. It is not so much a question of degree, but of one thing or another. Directly the proportion of oxygen falls short, the combustion changes.

26. The second process of combustion can be shown by a common fire, in which the products of the coal are best consumed, very little residue, and that a white ash only, being left. This is the slowest and most natural process of combustion, although a great amount of heat developed is lost to economy. A compromise between the close furnace and open fire is an improvement on either separately.

For instance, a donkey boiler keeping steam for

three steam winches will burn 14 cwt. of coal, and require a constant attendant. A large or main boiler will do the same work (with open fire-doors) on 30 per cent. less fuel and 60 per cent. less attendance, without clinker.

27. What is termed smoke is therefore the wasted products of the coal pure and simple. Every one knows that smoke condensed or precipitated is "soot," and that soot is unburnt carbon and oxygen, or condensed carbonic oxide. It is only a question of length of chimney to condense the smoke into soot altogether, and whether in the shape of soot, or of the caked crust of the hydrocarbons off the tea-kettle, both are solid burnable products of the coal.

Therefore, although, correctly speaking, we can lose nothing or create nothing, coal gases, whether exuded by evaporation or prematurely discharged by imperfect combustion, are to all intents and purposes lost to the present generation at least. Smoke proper is a product of a clear bright flame, as the top of gas flame, whereas popular "smoke" is carbon vapour, which is very different.

A. We require just sufficient air present at combustion to mate the liberated coal gases for perfect combustion, as at B, with no smoke; and as the combustion takes place on the top of the fire, if sufficient air cannot get upwards through caked or clinkered fuel, we must have holes in our fire-doors or furnace fronts to make up the deficiency; but if this be overdone, and too much air pass over the fire into the chimney, the loss will be as great as that of imperfect combustion, inasmuch as the air is cold and cold air cools the heating surfaces and destroys the draught. (See Fig. 4, L.)

B. Loss by insufficient or oversufficient quantity of air, 25 per cent.

Loss by radiation and in ash-pits, 8 per cent.

Loss by heated gases escaping up the chimney, 25 per cent.

Loss from first to last in the use of steam power, 90 per cent.

No matter how or what we plant, blight, weeds, and worm will assail.

We pay one man to build up, another to knock down.

The margin, however small, that the builder exceeds the knocker-down is termed progress.

28. One pound of coal:—

1 lb. coal uses 24 lb. air, which equals jointly 56 cubic feet.

100 lb. air $= 1315 \cdot 05$ cubic feet.

100 cubic feet air $= 7 \cdot 624$ lb.

1 lb. oxygen at $62°= 11 \cdot 88$ feet.

1 lb. nitrogen $= 13 \cdot 53$ feet.

	cubic feet.
23 lb. oxygen =	$273 \cdot 24$
77 lb. nitrogen =	$1041 \cdot 81$
100	$1315 \cdot 05$

Everything in nature is altered by change of temperature, so also is combination of coal gases.

29. A. No better example of these two orders of combustion need be sought than the candle and the ordinary paraffin lamp. When the wick is crusted, air-holes choked, or too much air admitted, very little manipulation of the wick will demonstrate unmistakably the perfect and imperfect forms of combustion, accompanied with their respective proportions of heat and light—the objects sought. It has also been here demonstrated that coal can be disposed of either with or without fresh air, both by evaporation and combustion, without the objects sought for being produced at all.

B. For example, in a common furnace a light bright fire, heavily charged and covered over with green coal, will have the flame choked out and dense smoke or wasted fuel the result, until flame is again recovered. Or if one part, say towards the back, had been left bright, this would have served the very important purpose of keeping the gases ignited, and, more important still, kept up the heat, just as one bright furnace consumes the smoke of another with a green fire. It would be more correct to say, keeps the green fire ignited until the flame is broken through or the fire itself, or part of it, has reached the point of ignition, 900° F.

30. § 16 shows how coal is consumed firstly by exudation up to 200°, then by evaporation (Fig. 4, E, F) up to 900°, or point of ignition.

Now, suppose we put a gas retort E, as representative of evaporation: a is a vessel made white hot by the fire b underneath, which is then charged with gas coal and closed. The dense vapour or gases disengaged by the heat are forced up the pipe *d* to a receiver, to be afterwards condensed and purified,

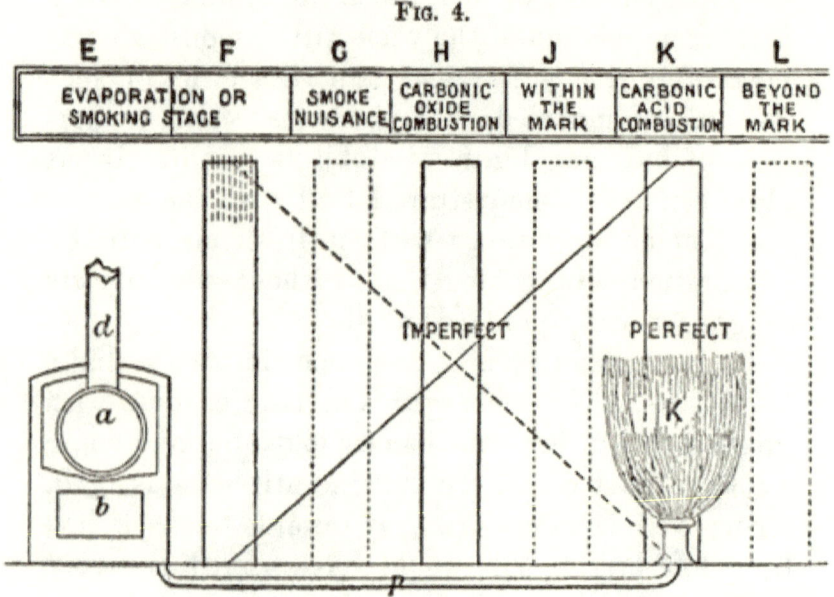

Fig. 4.

appearing afterwards as the luminous gas flame K, as representative of perfect combustion. The dotted line from top of F to bottom of K indicates that the evaporation is in full swing at F, converging to a point or nothing at K. The full line from top of K to bottom of F indicates that at K combustion is perfect, while convergence to a point shows there is no combustion there at all. Then putting K or perfect combustion as a focal point on the one hand, and F as a point on the other, any point between these must necessarily be a compromise between, and the position on the diagram definable by the proportions of its elements. (With J K see §§ 293-4.)

Therefore H being in the centre, or equality, has a proportion of carbon in addition to a proportion of perfect combustion, then adding one proportion of another we have exactly the elementary proportions of C, or carbonic oxide gas combustion; half-and-half, or imperfect combustion, as per diagram Fig. 3, which means 25 per cent. loss.

The position J is also characteristic of the dancing and intermittent stage of combustion. The position G may be put down as that of the smoke nuisance, while that of L, the proportion of air being in excess, means also 25 per cent. loss.

The ten dotted vertical lines at mouth of F represent these as each one pound, the intermediate lines as ounces.

Since there are no arbitrary conditions in the combining of the gases (1 ounce of carbon uniting with $2\frac{2}{3}$ ounces of oxygen just as readily as by pounds), therefore the sub-lines, if held as representing ounces or drachms, will just give the identical proportion as the primes, while the length of figure is not comparably altered.

If perfect combustion be 1 carbon and $2\frac{2}{3}$ oxygen, and imperfect combustion be 2 carbon and $2\frac{2}{3}$ oxygen, by multiplying the respective terms by 3 we have perfect combustion as 3 parts carbon and 8 parts oxygen, and imperfect combustion as 6 parts carbon and 8 parts oxygen, with the advantage of avoiding fractions.

31. All the coal products, tar, pitch, &c., excepting ashes, are formed into apparent smoke at E (Fig. 4), but the air being excluded there is no oxygen to support flame, therefore no combustion.

The coal consumed by this furnace underneath, in the production of heat only sufficient to disintegrate the coal or open up its interior to release the gases and secure them is 35 per cent. more than the gas coal itself.

The benefit we receive is got by hanging on to the gases during their escape through our hands. Smoke consuming is therefore a question of good combustion or bad. The pipe p from E to K underneath repeats the gasworks process.

J and K as opposites of each other (see §§ 293–4).

Very great assistance is given to the coal by breaking the lumps into small pieces. It is cheaper to break the lumps by hand before putting into the retort or furnace than by heat afterwards.

32. This would show that a saving would be effected by using small coal or coal dust. Theoretically, this would be a great smoke prevention desideratum, but it has up to the present been found a practical impossibility. With large coal the air is always at hand in the interstices, but the difficulty is to have heat enough to penetrate the lump. As coal gets smaller the same penetrative heating power is not required, but the interstices or air-spaces are also diminished; the difficulty of penetration is not removed, but changed. The small coal becomes a compact mass again, which cannot be penetrated therefore the matter is as broad as long.

33. If automatic firing could be practicable where finely fresh pulverised coal could fall without intermission on a thin bright fire, as dew on the ground, this would be combustion. It must then be plain that, in order to appreciate combustion, as nearly as possible the firing must be light and often, and with coal newly broken into pieces of the size of a man's hand. (See Firing, § 234.)

No better illustrations of combustion need be sought than the manipulation of paraffin lamps, and gas and candle flame (Fig. 3, C).

Generation of Steam.

34. The blowing of soap bubbles resembles much in appearance the formation of steam, the motive force in the one case being air ejected into the soapy water, the soap serving to magnify the phenomenon. The motive force in the other case is vaporised water forced from a hot plate. To watch the birth, life, and death of the one, is equal to reading a history of the other. While one emerges at the bottom with its charge of *mouth air*, another bursts, discharging its charge at the top, and so on, struggling into life and struggling out.

The smallest breath of wind is adversity to these frail figures, which are all intent upon reaching the top and *dying comfortably*, but, alas, many perish by the way from lack of substance. There is nothing lost, however, but *the number of their "mess"*; their remains going to increase the substance and thus prolong the lives of their mates: also a fact true of the struggles and vicissitudes through life of man and society. Again, as we have said, the soap produces the phenomena on a large scale, as the soapy matter is the substance, and is capable of great expansion and adhesion to the water, the soap being 8 per cent. of the framework itself encircling the air.

A. Again, take two tin pans, one half-full of soapy water, the other with pure water, and boil them; the size of the bubbles produced by the pure water will be very small indeed compared with the other, and if we take impregnated water, such as the sea, the bubbles will also be larger than that of the pure. Take the case now of the enclosed boiler generating steam, where the hot surface expels the water into vapour or steam. These bubbles, singly, or conjoined in texture like a honeycomb, leave the heated surface and seek the surface of the water, just as sparks fly upwards, undergoing many changes by the way, by currents and by decline of their heat, also head of water to overcome as well as distance to travel. On arrival at the surface they burst the enclosing vapour and take to the steam space. The water that constituted the framework of the bubbles will now be forced by the crowd of other bubbles discharging to the side, where it will again descend to the hot plate for another charge of steam. Many of these ascending bubbles, however, perish by the way; the atom of heat they were conveying is just imparted to the water. This framework of the bubbles is just the colder and denser part of the water; the purer the water the smaller these globules, and the greater the heat the greater the ebullition or multiplication of them, or generation of steam.

35. Now we must bear in mind that, notwithstanding the impurity of the boiler water, whether muddy or highly impregnated sea water, the contents of these bubbles discharged at the surface are purity itself, pure distilled water vapour, and that, with the fact of the steam impinging on and condensing constantly over the whole surface, added to the fact that fresh water keeps to the surface, is proof enough that no matter what the density of the boiler may be the surface of the water is always fresh.

Conformably with this we find that the colder and denser atoms of water descending from the surface find their way to the hottest surfaces, where a levy is continually going on for fresh water to make steam; these atoms that have just descended will again be made up (minus the solids) and fulfil their mission as before.

This steam manufacturing process is more simple with pure water than with impregnated, in which case the ebullition is fitful, and with the ebullition stopped or impaired, the generating of steam is attended with risk of the solids in the water adhering to the heating surfaces, causing

INCRUSTATION OR SCALE.

36. In view, therefore, of the great risks accruing from scale, the necessity of preventing it is a stern

one to all having to do with steam boilers, whether using water salt or fresh. (See §§ 164, 175, C, B, D, 177, 161, 158.)

Besides the counsel given there, we can also take counsel of our own domestic experience, in the way of cooking dense fluids, such as gruel, milk, &c.: we find that on boiling milk in one tin pan, and water in another, we have boiled the clean water all right; but in the case of the milk, it has all adhered to the bottom, leaving a mixture of whey on the top. Now, the old lady will tell us we ought to stir milk with a spoon when we boil it, or it will sit to the bottom; so will gruel, worse than milk, and she will add with some severity, "of course." Now this stirring was just circulation of the water, baffling it in its endeavour to settle by stirring. We find the old woman also, in such a case, uses a slow part of the fire in addition to the stirring, as she has it strongly by tradition, with the proofs of experience, that if she does not stir, and boil slowly, the *stuff* she is boiling will *sit to* the bottom of her pot. The fire goes on heating, the solder melts, and the bottom drops out. Now all this is in consequence of this attachment of the solids for the hot plate, which, if circumvented, makes all the difference between scale and no scale, or, circulation and none. (See § 162.)

37. If salt water, for example, be introduced upon a hot plate, as in the sketch (Fig. 5), the water at the coldest part divides into large bubbles, and, as they travel on to the hotter parts, they burst. In the

case of pure water, this may be carried on long enough without any deposit being left on the plate. But with say dense sea or other impregnated waters, we will have a deposit of solid matter in proportion to the density, and in structure conformable to the process. We introduce dense water from the pipe w; P H is the hot end of the plate; L is a spirit-lamp.

Now we all know that water will not remain upon a hot surface, as with a cherry red heat in this case, we find there is quite a clear space between the drop

Fig. 5.

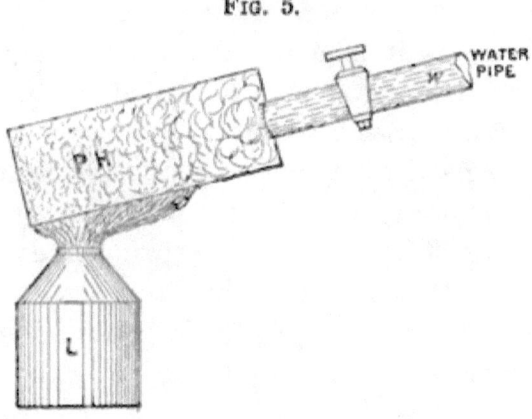

of water and the hot plate, because the drop is borne up by the force of the generating steam, but the solid matter in the drop is immediately deposited, while the pure or liquid part vaporises into steam. We can have the deposit on the plate going on so long as we keep on the water, until it gets so thick as to interrupt the heat, when we are stopped, whereas with pure water there would be no limit.

Now, as with the experiment, so with the reality. Over an intensely heated part, such as a furnace crown, and with highly impregnated water, its structure resembles a sponge, or honeycomb, which, as time goes on, increases its substance, thus decreasing the area of the cooling part, and when this cooling part is so diminished as to be unable to cope with the depositing tendency of the framework part, deposit of scale is a matter of course.

38. Some boilers, owing to better construction for circulation, or more capacity for generating steam, with other inherent qualities, may not take on scale even with dense water. A new boiler digests the solids best, as it is new. After once acquiring a scale, the fate of a boiler is sealed, *it is a " salt-seasoned job."* (See § 105.)

39. The particular times to acquire scale are, when a boiler has been blown down hot; suddenly stopped and blowing off steam; keeping steam handy (§ 258); banked fires (§ 274), when ebullition stops.

After scale has once got footing on a boiler and is properly formed it can only be properly removed by chipping. It is impracticable to dissolve chalk, gypsum, alabaster, marble, &c., without injury, or great risk of injury, to the boiler itself.

40. The loss owing to scale and its consequences is much greater with marine boilers, owing to the complexity of their internal parts, which is an enemy to the only natural preventive of scale—circulation, and scale formation may be put down as a property inherent in all boilers more or less. It is incumbent

on all in charge of boilers, notwithstanding, to use all fair means to prevent its deposit by discretion in the use of water (§§ 8–12), and water and fire management (§ 234). Failing this, the use of chemicals, like that of medicine to the human subject, has to be resorted to.

41. As salts of lime is the trouble referred to being deposited in the boiler, it has been operated upon by Clark and others by endeavouring to purify the water before it enters the boiler. It is certainly better to catch the salt before it enters the boiler than to have to chip it off the intricate and inaccessible parts of a boiler. But to medicate the great amount of water used by a H.P. boiler sufficiently to neutralise the lime solids would be out of the question, generally speaking. However, the plan of adding lime to water impregnated with it, which is the same in principle with setting a thief to catch a thief, has something to be said in its favour.

A. There is a story of an ant who bearing home a burden of food failed entirely after many attempts to climb a bank. At length it gave it up for the time being and rested, partaking during the time of a hearty meal off its own burden, which, of course, lightened it. At the same time it added strength to the individual. Thus with lightened burden, rest, and refreshment, the task was accomplished with ease.

B. A shrewd contractor when contracting to fill up a hollow will always look out for a hill requiring removal to fill it up with, just as in the construction of a railway line, hill and hollow have to make common cause for a mean level.

42. Upon such lines we must go in the treatment of water for boiler purposes, viz. we must, like the ant and the contractor, call one thing to the aid of another whereby to kill two birds with one stone. We must have a Roland for our Oliver—a fair exchange. Even the savage knows that barter gives a mutual benefit to both parties if they bargain with their eyes open, which is correct.

Then, to dispose of the solids in water as chemical substances we use other substances producing such chemical exchanges as are desired. No compound whatever ought to be put into a boiler without its character being known. To put any simple substance into a boiler is a compound matter. It will always do two things, possibly what is desired of it by the first combination, and possibly the reverse by the reaction. That is, we may have good results as to prevention of scale in the first instance, but the reaction may be very detrimental to the boiler shell. Therefore, it must be made a point to understand the tendencies and follow them up to the end, and the great desideratum of chemical exchange must always be looked out for. If there is no chemical action and the substance remains on the heating surfaces inert, the risk of burning the boiler is very great.

WATER ANALYSIS.

43. A (Fig. 6) is a tube filled with sea water from 1 to 0, and divided into 33 parts. 1 at the bottom shows the amount of solid matter in proportion, or $\frac{1}{33}$, left after the evaporation of the water; 2 shows the additional deposit on a second charge of the tube; and 3 a third charge, corresponding with the register of a salinometer (see § 104), or $\frac{1}{33}$, $\frac{2}{33}$, $\frac{3}{33}$.

FIG. 6

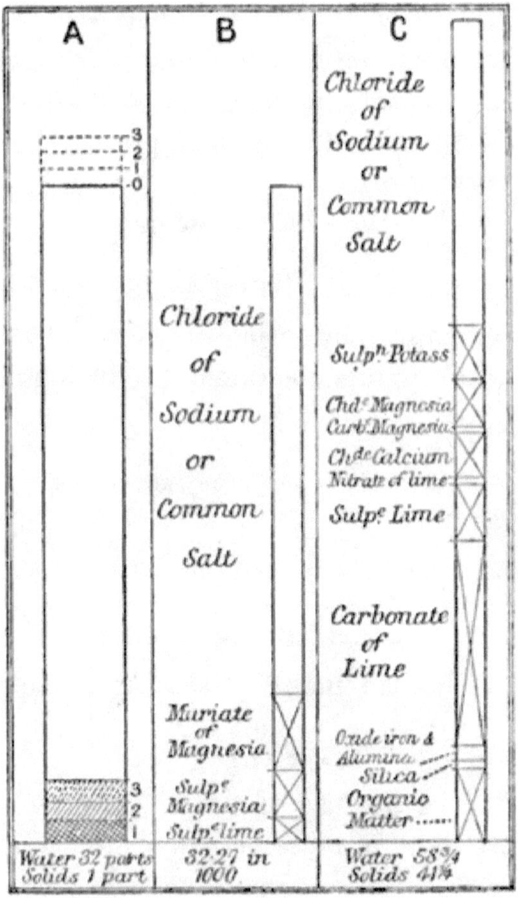

A

B — Chloride of Sodium or Common Salt
Muriate of Magnesia
Sulp.ᵉ Magnesia
Sulp.ᵉ lime

32·27 in 1000

C — Chloride of Sodium or Common Salt
Sulp.ʰ Potass
Chd.ᵉ Magnesia
Carb.ᵗ Magnesia
Ch.ᵈᵉ Calcium
Nitrate of lime
Sulp.ᵉ Lime
Carbonate of Lime
Oxde iron & Alumina
Silica
Organic Matter

Water 32 parts Solids 1 part

Water 58¾ Solids 4¼

D

B illustrates similarly the solid matter contained in one charge of the tube, on a larger scale for illustration, and corresponding with the analysis of Faraday. It shows about three-fourths of the solids to be common salt, the remainder muriate and sulphate of magnesia, and sulphate of lime, in various proportions.

C illustrates Dr. R. D. Thompson's analysis of Thames water. $\frac{1}{33}$rd, 1000, and 100 signify alike.

Elements of Nature.

44. There are only 67 substances in nature altogether, 52 of which are metals. We will only deal with those affecting the subject presently, or likely to do so further on.

In all combinations of the elements of nature

Heat

is the prime mover, by acting on the solid substances of which the earth is composed, producing acids and alkalies.

Acids	*Alkalies*
Are sour and corrosive substances, and are put down as proved acids when they turn *blue* litmus solution *red*.	Are neither sour nor corrosive, but have the power of neutralising acids, and also of turning *red* litmus solution *blue*.

Acids and Alkalies

combine with each other, producing a neutral body called a

Salt.

Therefore in the use of water for boiler purposes we can test it by a piece of paper dipped in litmus solution and put in the water. If it contains lime or other alkali, the paper will be turned red; if bearing acids, blue. They combine in the boiler and produce a salt all over the heating surfaces as a scale.

ELEMENTARY SUBSTANCES.

45. Metals:—

Aluminum (Al = 27).	A metal extracted from clay and alum crystals; produces alumina, the earth of clay.
Calcium (Ca = 40) ..	A metal extracted from lime, gypsum, alabaster. Lime, marble, chalk, &c., are all calcareous.
Magnesium (Mg = 24).	A soft metal. When burnt it leaves magnesia, a white powder.
Potassium (K = 39)..	A metal contained in potash and in potash salts.
Sodium (Na = 23) ..	Is the metal of the soda salts, very suffocating, yellowish gas.
Mercury (Hg = 200).	Most sensitive metal.
Copper (Cu = 63). ..	
Lead (Pb = 207)	Softest metal.
Iron (Fe = 56)	Commonest of metals.

Each metal has its ore, in which it is carefully protected and secreted, similarly to the seed in fruit.

46. Non-metals:—

Oxygen (O = 16) ..	An invisible, tasteless, and colourless gas	Air.
Hydrogen (H = 1) ..	Also invisible, tasteless, and colourless gas	Water.
Carbon (C = 12) ..	A solid element existing in coal, diamonds, blacklead.	Coal.
Chlorine (Cl = 35) ..	A strong-smelling and poisonous yellow gas, got in salt	Bleaching powder.

Sulphur (S = 32) ..	A solid found in volcanic districts	Brimstone.
Phosphorus (P = 31) .	Got from the ash of burnt bones, red and yellow. The shining substance of dead fish.	
Silicon (Si = 28) ..	A black crystalline substance —quartz, flint, rock-crystals.	Flint.
Nitrogen (N = 14) ..	An invisible, tasteless, and colourless gas	Nitre.

COMPOUNDS.

47.

CHEMICAL NAME. Gas or Metal Bases.	COMPOSITION. Salts.	COMMON NAME.
Chloride sodium ..	Issue of Chlorine and sodium ..	Common salt
Sulphate potass ..	,, Sulphur and potash ..	Brimstone and potash
Chloride magnesium .	,, Chlorine and magnesium	Metal and base
Carbonate magnesia .	,, Carbon and magnesia ..	Acid and base
Chloride calcium ..	,, Chlorine and calcium ..	Metal and base
Nitrate lime	,, Nitre and lime	Saltpetre and lime
Sulphate lime	,, Sulphur and lime ..	Acid and base
Carbonate lime ..	,, Carbon and lime	Acid and lime
Oxide iron	,, Oxygen and iron	Rust of iron
Alumina	—	Oxide of earth and clay
Silica	—	Quartz, sand, rock, &c.
Organic matter	—	Decomposed vegetable-life
Sulphate soda	,, Sulphur and soda	—
Sulphide lead	Galena	Lead ore
Hydrochloric acid ..	Hydrogen and chlorine	Acid and base
Sulphate copper ..	Sulphuric acid and copper ..	Bluestone

48.

Calcium chloride, soluble,	Sodium carbonate, soluble in water,

Exchange, and become

Calcium carbonate, chalk, insoluble in water.	Sodium chloride, common salt, soluble in water.

49.

Sodium chloride (common salt) and sulphuric acid

Exchange, and become

Sodium sulphate (Glauber's salts) and Hydrochloric gas.

50. Muriatic acid and chloride of magnesia combined is a very destructive agent.

51. Carbonate of lime and magnesia, and such earthy salts, are kept in solution by the free carbonic acid gas which the water contains. By boiling, the gas is expelled and the salts precipitated.

52. When spring and well waters are boiled the free carbonic acid is driven off, and the carbonates, deprived of their solvent, are precipitated in a finely crystallised form.

53. With sulphurous coal and great heat the boiler shell is liable to suffer, the sulphur combining with the iron producing bi-sulphuret of iron.

COMBINING QUANTITIES.

54. The capital letter placed beside the element is just an abbreviation or initial, as a person uses initials for shortness, and the figures beside the letter represent the fixed proportions by weight in which the substance combines with others. Let us take the *metal* aluminum and the *ore* alumina.

Again, aluminum $Al = 27$, means that 27 parts is the combining weight with other substances, or the *individuality* of the element aluminum, and so on with the others.

55.

NH	form	Ammonia.	NHO	form	Nitric acid.
SO	,,	Sulphuric acid or oil of vitriol.	CHNSO	form	Coal.
			H_2O	form	Water.
SiO	,,	Silica.	ZnO	,,	Zinc oxide.
SMg	,,	Epsom salts.	HgO	,,	Oxide mercury.
NO	,,	Common air.	KNO	,,	Nitre.
CaO	,,	Calcium oxide.			

Sulph*ates, ides,* and *ets,* &c... Compounds pertaining to or issuing from sulphur.
Carbon*ates* .. .. Compounds pertaining to or issuing from carbon.
Nit*rates* .. .. Compounds pertaining to or issuing from nitrogen or nitre.
Chlor*ates, ides,* &c. .. .. Compounds pertaining to or issuing from chlorine.
Silic*ates* .. .. Compounds pertaining to or issuing from silicon.
Phosph*ates* .. .. Compounds pertaining to or issuing from phosphorus.
Ox*ides* .. .. Compounds pertaining to or issuing from oxygen.
Hyd*rates*.. .. .. Compounds pertaining to or issuing from hydrogen.

56. The prefixes and affixes denote the relations of the compound, the symbols or initials the individuality or combining proportions.

It is true throughout nature that chemical combinations take place more readily between bodies which least resemble each other.

Scale Analysis.

57. Sulphate of lime, the worst of boiler scale, being 86 per cent. of that found in boilers, is the same substance as gypsum or plaster of Paris and alabaster, and is held in solution in both fresh and sea water. It adheres to the heating surfaces, forming a hard tenacious crust, in course of time becoming rock itself. It deposits as the steam rises, up to 75 lbs. pressure; what is not then deposited is held in suspension, which condition most favours the formation of scale on heating surfaces.

Scale Analysis.

Fig. 7.

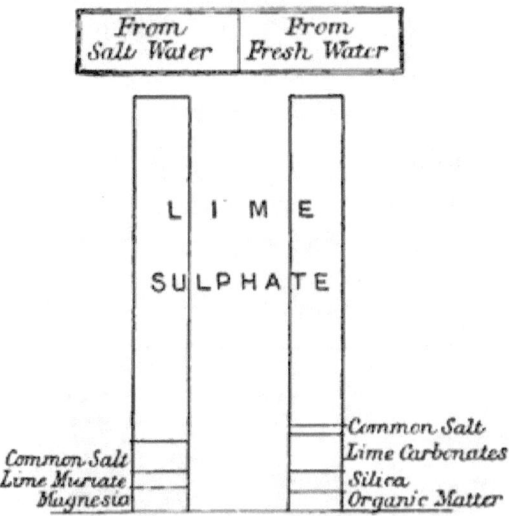

58. Carbonate of lime is the same substance as selenite. Chalk, marble, and limestone are mostly all deposited before steam has been got up to 60 or 70 lbs. pressure. They only form about 7 or 8 per cent. of boiler scale. When depositing they find their way to the bottom in the eddy of the shell or other quiet place.

59. Carbonate of magnesia deposits partly along with carbonate of lime, both in marine and in land boilers. It forms along with the other about the feed and other orifices.

60. Common salt, a large ingredient in sea water, only deposits when at great density. It is seldom to be found in ordinary boiler scale, except in minute quantities. Nitrate of lime, very nearly common salt, probably forms an intermediary flux, facilitating

the deposit of common salt. Hydrochloric acid is an active corrosive inside a boiler; sulphuretted hydrogen, an active corrosive in combustion chamber.

WATER ANALYSIS.

61. The following table, by Berthier, shows the order and proportionate deposition of solids upon concentration by boiling off steam from sea water:—

EACH 100 PARTS.

Saline Matter.	A Sea water sp. gr. 1·0278 or One 33rd.	B Sea water concentrated sp. gr. 1·140 or Five 33rds.	C Sea water concentrated sp. gr. 1·220 or Eight 33rds.	D Sea water concentrated, salt on point of deposition or Twelve 33rds.
Chloride of sodium	2·50	16·00	25·50	20·80
Chloride of magnesium	0·35	0·46	1·07	4·85
Sulphate of magnesia	0·58	0·80	1·48	9·50
Carbonate of lime and magnesia	0·02	0·00	0·00	0·00
Sulphate of lime	0·01	0·30	0·00	0·00
Sulphate of soda	0·00	2·65	2·81	0·00
Water	95·54	79·79	69·14	64·85
	100	100	100	100

62. Thus when sea water is boiled down so that the water is only about 65 per cent., all the salts of lime and magnesia, and the whole of the sulphate of soda formed, and a large proportion of the common salt deposited at D, the boiling point is high. The extraordinary density of C and D must not be taken as showing the capacity of water to carry solids in solution, but as showing the chemical combinations of

solid substances inside of a boiler at work. The popular theory of mechanical combinations is clearly disproved.

DENSITY.
Combination of Solids.

The first column shows sea water pure and simple; its ordinary density, say $\frac{1}{33}$, the second increased by about 5, then 8, then 12, approximately. Taking the first column as ordinary sea water, we note there is no sulphate of soda, and that there is one part carbonate of lime and magnesia. Well, in the next column this has disappeared—very curious! and we have now $2\cdot 65$ of sulphate of soda. Now, where has this come from? In the next column the sulphate of lime has disappeared also. In the last column D the sulphate of soda has disappeared as mysteriously as it came. Now there has been nothing added or taken from the water all the time, save what has been done by the evaporation.

63. The disproportion between A and D will show the fallacy of mechanically calculated results, as well as what is otherwise going on in the boiler. Common salt, however, is an exception, as it maintains its individuality very nearly to its depositing. We find a difference of one-third between the real and calculated results, which means that the salt is depositing, and that sulphate of magnesia gains and loses in great disproportion, which shows that it is an active combining agent. The disappearance of carbonate

of lime and carbonate of magnesia at an early stage indicates that these salts are deposited at a low temperature, or that the carbonate of lime is deposited and the magnesia resumed in process, or *vice versâ*.

64. Now looking along the figures we find the extraordinary increase of sulphate of lime between A and B, and its sudden disappearance afterwards shows that at first it was only in formation, and that it increases very rapidly and deposits similarly. We find also that sulphate of soda has been manufactured inside the boiler entirely by combination of the other substances, and then goes as it came. Now as we know that these substances maintain the integrity of their proportions, that is, their *combining weights*, the elements of the water included, therefore while the evaporation is going on there will be continual change. And assuming nothing has escaped from the boiler but steam, no solids have been lost, nothing can be lost, and nothing created.

Therefore the column D will represent the same elementary substances of the solids as A. The disparity is the question of chemical *exchanges* and combinations.

65. The extraordinary saturation of D must not be held as a criterion in the question of density, as it is not the amount of solids, but their character and respective proportions, that is the matter of anxiety with the engineer.

66. Sea water at the density of D decomposes, in consequence of the loss of its fixed air (§ 179).

Specific Gravity and Density.

67. The illustrated analysis (Fig. 6) is intended to show the disparity of solids, as may be measured with the eye (Fig. 6, B), and has had a little more lime added to it to conform with the reading of the salinometer as at Fig. 6, A; the table is therefore here inserted in its integrity.

Faraday.

	Parts.
Chloride of sodium	25·762
Muriate of magnesia	3·282
Sulphate of magnesia	2·212
Sulphate of lime	1·013
	32·269

Specific gravity of pure water .. .. 1·000 62° F.
     ,,   ,,   ,, sea water .. .. 1·027 ,,
Weight of pure water .. 62½ lbs. per cubic foot.
     ,,   sea water .. 64·1416 lbs. per cubic foot.

The density of sea water is put down at 33 parts in 100 as a fair average.

	Density.	Boils.	Specific Gravity.	Ounces of Solids per Gallon.
Pure water	0	212°	1	
Sea water	$\frac{1}{33}$	213·2°	1·029	4
	$\frac{2}{33}$	214·4°	1·058	8
Must not exceed this	$\frac{3}{33}$	215·5°	1·087	12
	$\frac{4}{33}$	216·7°	1·116	16
	$\frac{5}{33}$	217·9°	1·145	20

68. From the foregoing table it will be seen that the quantities increase in proportion. We speak of our boiler water as being at 1·2 or $\frac{3}{33}$ *density*,

whereas scientific writers put it down as so much *specific gravity*, which practically puts these as synonymous terms, and such is the teaching of the table. But there is a difference. See § 82, salinometer. Besides, we have often the density of water given us only in this language. The following is a table of specific gravities of sea water at 60° F. in various parts of the world, as ascertained by Dr. Marcet:—

Arctic Ocean	1·0266	Sea of Marmora	..	1·0191
Northern Hemisphere	1·0282	Black Sea ..	..	1·0141
Equator	1·0277	White Sea ..	..	1·0190
Southern Hemisphere	1·0288	Baltic ..	..	1·0152
Yellow Sea..	1·0229	Iceland	..	1·0005
Mediterranean ..	1·0293	Dead Sea ..	..	1·2110

Density and Boiling Points.

69. Again, we always speak of the boiling point of water as 212° as a rule, irrespective of its density or purity. Now we have just read in Table B that pure water boils at 212°, and common sea water at 213·2°, and at $\frac{3}{33}$ it is 215·5°. Now two degrees of the thermometer scale is a very small space to work on, if testing our density of boilers by that instrument—so small that it is unsatisfactory. These instruments are not got so nicely adjusted. Now all engineers acquainted with this mode of testing the water also know, if they have not forgotten, that the height of the barometer should be consulted. Now it is a fact that very little importance is attached to the atmospheric pressure at all. In most cases there is no barometer below, and that is enough.

Barometer at Level of Sea.	Boiling Point. Fahr.	Barometer at Level of Sea.	Boiling Point. Fahr.
27 inches	206·96°	29½ inches	211·29°
27½ ,,	207·84°	30 ,,	212°
28 ,,	208·69°	30½ ,,	212·79°
28½ ,,	209·55°	31 ,,	213·57°
29 ,,	210·38°		

Well, the difference between the highest and lowest point of the above table is 6·61° F., and throwing out one top and bottom, the difference is yet 4·95°, more than double the range of our index altogether, and if the boiler was at $\frac{2}{33}$, or less, when we tried, this makes it worse, and is plain enough to show that unless the thermometer be correct, and the barometer carefully noted, very ridiculous results would be obtained, and probably, if depended on, perilous consequences too.

SCALE FORMATION.

70. We will find on looking over the analysis (§ 61, A), that chloride of sodium or common salt is the prevailing solid in sea water, being about two-thirds of the whole. Then follows chloride, and also sulphate of magnesia, then sulphate and carbonate of lime. All authorities on the subject, scientific as well as practical, put down sulphate of lime as the great scourge of steam boilers. This decision is arrived at from the fact that that solid constitutes between 80 and 90 per cent. of all the scale formed in steam boilers both land and marine, and it is also the most intractable and obdurate to remove of any

scale that could be formed. Therefore the one great object aimed at is its prevention, which is as yet unattained.

71. Prevention suggests either the extraction of the noxious solids from the supply water, or the intelligent manipulation of the boiler itself while working. The problem is, Which will in the end be most successful? Extracting or otherwise decomposing these solids before entering, or, after having entered the boiler, being a chemical matter, places the subject in the hands of the chemists and druggists, as well as quacks, resulting in no end of "cures" for boiler scale. The management and treatment of boilers by those in charge, many of whom are also quacks, is just as varied and complicated as the circumstances on the one hand, or scale compositions on the other. We will review several modes of boiler treatment by those in charge of boilers.

Blowing Practices.

72. One school says, "Blow the boiler when you can; keep her fresh," believing that the nearer the boiler can be kept to the original density, the less chance of scale with salt water, and accounting for formed scale thus:—That it was lime got in river water, and that there being no salt in the scale to speak of, hence their success in freshening their water when they had a chance.

73. Another says, "Never blow a boiler; as the more blowing the more scale, and the less blowing

the less scale," and this irrespective of circumstances (§§ 92–5).

74. Another sect are in favour of having a thin scale all over the internal surfaces, the reason being to protect the boiler from injury; and if this cannot be got, they put slaked lime inside, with sufficient heat to get the lime to adhere, thus forming a scale. (§ 102).

75. Others stick rigidly to the teaching of the salinometer reading, and blow at $2\frac{1}{2}$-33rds density, without knowing that a boiler might be burst with scale whose density never exceeded $\frac{2}{33}$rds. Some are continuously dosing their boilers with astringents to prevent scale, which runs the risk of "pitting." Many run their boilers on the mechanical principle of dead reckoning for the density.

76. Others claim to be able to keep a boiler clean, or scaled as desired. Others claim to have a mode of treatment of their own which they keep a secret; two-thirds of those in charge of boilers have the fallacious idea that the salinometer represents all the solids in the sea water, and that with the instrument at 0, there are no solid ingredients in the water at all, without thinking of any changes by expansion of the water, or heat, or that there is any other scale deposit but that common to all boilers (§ 91), and that that is composed of common salt, the existence of sulphate or carbonate of lime scale being next to unknown. Amongst the nondescript class of engineers especially there are some very extraordinary ideas in this respect, as well as modes of treatment, foremost

being that of freshening the water (§ 72). We endeavour here, by the aid of facts and figures, to disprove that common fallacy without overstepping the mark (§ 73).

Blowing out Water for Freshening.

77. Firstly, we advert to the analysis of sea water (Fig. 6, B), also that of fresh water also impregnated (Fig. 6, C), and compare these together, and then with the analysis of boiler scale (§ 56). We take the best known form of impregnated water, namely sea water, to begin with, which is as 32 of water to 1 of lime; as this water vaporises into steam, the solids remain behind, or, in other words, the water is boiled off and the lime is left. It is therefore plain, that with every 32 gallons of water we take into the boiler to evaporate away, we take also 1 gallon of solid lime salts to remain.

78. If we wish the boiler not to exceed $\frac{2}{33}$rds density, we must blow out the water thus:—

	Galls.	33rds density.		
For	1 at	0	{ evaporation	= 2 galls. put out with
We blow out	1 at	2	for exchange }	2 quotas of lime.
Take in	2 at	1	from the sea	= 2 galls. taken in with 2 quotas of lime.

If not to exceed $\frac{3}{33}$rds density, we must blow out the water thus:—

	Galls.	33rds.		
For	2 at	0	{ evaporation	= 2 galls. of water and 3
Blow out	1 at	3	for exchange }	quotas of lime put out.
Take in	3 at	1	from the sea	= 3 galls. of water and 3 quotas of lime taken in.

Freshening.

Density.	Evaporate.	Blow out.	Take in.
2-33rds	1	1	2
2¼ ,,	1¼	1	2¼
2½ ,,	1½	1	2½
2¾ ,,	1¾	1	2¾
3 ,,	2	1	3
2-33rds	6	6	12
2¼ ,,	6	4¾	10¾
2½ ,,	6	4	10
2¾ ,,	6	3½	9½
3 ,,	6	3	9
3¼ ,,	6	2¾	8¾
3½ ,,	6	2½	8½
3¾ ,,	6	2¼	8¼
4 ,,	6	2	8
4¼ ,,	6	1⅞	7⅞

Therefore, to maintain $\frac{2}{33}$ density, we blow out the same amount as we evaporate, and draw from the sea a quantity equal to both. One-half of the total water is blown away. For $\frac{3}{33}$ density we only blow out an amount equal to one-half of the evaporation, and just one-third of the amount taken from the sea.

As these facts reduced to figures cannot be disputed, the fallacy of working boilers fresh to avoid scale is clearly proved.

It is easy to remember that, to keep a boiler on salt water running at 1½-33rds, we must blow out ¾ of the feed water; at $\frac{2}{33}$, one-half of the feed water; $\frac{3}{33}$, one-third; $\frac{4}{33}$, one-fourth; and so on.

These tables refer to evaporation only. If there be leakage from tubes, or otherwise, of water, then it may be unnecessary to blow any water out. But the mischief lies in taking water in.

Specific Gravity and Density.

79. Specific gravity is the difference of weight between any body or substance, and pure water bulk for bulk. If the body is lighter, it will float, as will wood, or any hollow vessel; if heavier, such as a stone, it will sink, the specific gravity of the stone being greater. The specific gravity of sea water is variously estimated by different authorities; the difference between them, however, does not sensibly affect the matter, at least so far as we are concerned. On referring to Berthier's table (§ 61), we find the specific gravity of sea water $1 \cdot 0278$, and containing $3 \cdot 46$ solids and $96 \cdot 54$ of pure water, and on referring to the table (§ 67), we have $\frac{1}{33}$ of density, as equal to $1 \cdot 0278$ specific gravity, the former being the practical or engine-room phraseology, the latter the equivalent in scientific.* Therefore, if we assume that as $\cdot 0277$ is the difference between fresh and sea water, we have

Density.		Specific Gravity.
0		0
$\frac{1}{33}$ of solids	as	$\cdot 0277$ of solids
		2
$\frac{2}{33}$ equals 2 of gravity ..	..	or $\cdot 0554$, and so on,

which is sufficient for our purpose.

80. Density per salinometer when under steam is generally adhered to as an infallible index to the progress of scale in the boiler. Now when we com-

* Berthier and Faraday slightly differ.

pare the analysis tables, and other testimony of scientific authorities on the subject, we find that with high-pressure boilers the bulk of the solids is deposited before steam is up, carbonate of lime at 212°, and sulphate of lime at 280°, the temperature of our steam being 300°. Therefore, when we speak of a boiler being at $\frac{2}{33}$ density, what is meant by it, or what does it really indicate? If $\frac{1}{33}$ means 100 per cent. of the solids in sea water, and if 86 per cent. of this $\frac{1}{33}$ be deposited on the first getting-up of steam, the remainder will be 14 per cent. of this $\frac{1}{33}$ or $\frac{1}{206}$, assuming this theory to be correct. Then why does the salinometer say $\frac{1}{33}$? Again:—

DEPOSITION OF SOLIDS.

81. Sea water as drawn from the sea at 60° = 100 per cent.,

„ „ deposits carbonate of lime at 190°,

„ „ „ sulphate of lime at 280°;

and at 280°, 86 per cent. of the solid matter is deposited. 100 per cent. = $\frac{1}{33}$, and 100 − 86 = 14. Therefore 14 per cent. of $\frac{1}{33}$ or $\frac{1}{206}$ is the theoretical density.

Then why is the water so pre-eminently salt to the taste? And what is meant by density by salinometer? And if the major part of the solids taken in be common salt, wherefore comes the major part of the solids left as scale to be lime? Where has all the salt gone?

The saltness of sea water to the taste is owing to

E 2

the presence of common salt, which is about two-thirds of the whole. When it gets concentrated in a boiler by evaporation its saltness will be increased in proportion, and this will be shown by the salinometer.

SALINOMETER.

82. Density by salinometer is a very vague term. It means, however, to a great number of engineers the aggregate amount of solids taken into the boiler from first to last (minus what may have been blown out with the water), and they believe that the disposition of the boiler to take on scale will be nearly in ratio with the reading on the stem. Others have the belief that no deposit will take place under $\frac{3}{33}$, and freshen the boiler at $2\frac{1}{2}$ accordingly.

83. The salinometer only indicates the solids actually in solution with the water as part and parcel thereof. As we are always well under weigh before trying the instrument, the salts of lime are long ere this deposited, or held in suspension only. These salts at no time affect the salinometer, as it is not available under, say, 300° F. At any rate these are deposited before it can be used.

CHARACTER OF SCALE.

84. Of the solids admitted into the boiler with sea water, salts of lime are deposited; while common salt with a small addition of those of magnesia is held in solution, although concentrated as three to one, and will be discharged accordingly with the

water, unless extraordinary circumstances have provoked its previous deposit (§§ 92, 307). The deposition of the salts of lime is put down to elevation of temperature. But no doubt the expansion of the water itself would also account for it mechanically (§ 209). In good circulating and properly timed boilers these salts are found in the bottom as a white, soft, segregated mud. In others this is only partially accomplished, a large percentage adhering to the heating surfaces, especially the tubes, as common scale.

85. An analysis of boiler scale, and analysis of the feed water, are, as shown at §§ 57, 61, very different. So also is the character of scale itself taken from different parts of a boiler—the mean duration of the deposit, circulation of water at the part, as well as the intensity or otherwise of the heat. Tube scale, as well as that on combustion chambers, is always much softer than that deposited on furnace crowns or combustion-chamber tops, as these surfaces, owing to the force of ebullition, can pretty well keep clear of common scale, but generally secure a scale during the blowing-out of the dense water, chiefly of nitrates of lime and magnesia, as well as common salt. This is easily removed, being often cracked. Patches of scale on these surfaces lead to accumulation, which is very dangerous (§ 203). So also do waste or old bags that may be left in.

86. It is not yet plain what $\frac{1}{33}$ and 4 ounces per gallon on salinometer-stem refer to, or if the instrument is more available in ordinary fresh waters. That

salinometer was marked thus before it was discovered that the water solids were deposited at so early a stage. It can only tell of present density, without reference to any solids previously in solution. The reading is therefore wrong, but it is easily translated: thus, $3\tfrac{1}{33}$ to mean one strength of brine; $\tfrac{2\tfrac{1}{2}}{33}$ two-and-a-half brines, or concentrations of the water. Used in the same way, it will be no more available for fresh water unless graded and loaded suitably for lower temperatures.

Deposition of Solids.

87. Salts of lime either suspended or deposited are to be dreaded, and the sooner deposited matter is blown out the better. Suppose a boiler has just been filled from the sea, and steam got up to 80 lb., remaining so, say, six hours. All the carbonate and sulphate of lime is deposited, excepting what may be still held suspended. Say we now bank fires after boiling ceases, all the internal parts will be coated over, preferably the heating surfaces, to which the lime adheres a fixture. As the boiler is allowed to cool, the water will resume its lime conformably with the period of induration, that is, a boiler that has been steaming a number of days with the same water will yield up more of the solid matter, under the same blowing and cooling-down circumstances, than with steam up only six hours, as the case in question, which is all the difference between the lime salts being deposited and being resumed by the water,

which will at its normal temperature naturally resume its normal condition. This resuming, similarly to the depositing, gives our heating surfaces another coating.

88. Raising and letting down of steam gives two coats of lime each time, or one coat every agitation. Lying under banked fires is a very insidious practice for producing scale on the tubes; worse still is that order "keep steam handy," which means blowing off steam and loss of water. *Steam handy* is therefore a compound evil (§ 258). Irregular treatment in firing or easing fires, or anything tending to priming or agitation of the water, all conduce to scale.

89. With marine boilers the greatest sufferers from scale are passenger and other river steamers using impregnated water, by the frequent stoppages and waste of steam, with no time in port to allow boilers to cool down, but blowing out and baking the scale on the tubes. Ocean travellers, on the other hand, have all the circumstances in their favour, no blowing off steam, no loss of feed water, no tearing up of glands reversing engines, and no leakage therefrom, no racing against other vessels, and being periodically cleaned, keep themselves clean. New metallic heating surfaces retain the property of keeping free of scale for a long time; but having once got a scale, means very nearly always to have it. In tubed boilers the tubes are the greatest annoyance for taking on scale, and after getting a footing it increases very rapidly.

90. Scaled tubes, although not a dangerous con-

dition, is a fatal one to proper consumption and general economy. Of course increase of scale means increase of fuel. But that is not the worst part of it, for in proportion as the tubes add to their diameter by scale, so is the circulation of the water, and generation of steam obviously impeded (§§ 150–1).

Salt Deposit.

91. (§ 99) So much for common scale; now we come to speak of the second deposit or that of common salt. The most favourable conditions are with tubes and other parts scaled up, such as referred to; the feed water not returned; firing hard; coals used giving out most of their heat locally with no flame. These conditions place the furnaces in great danger, as the deposit of common salt is thereby induced (§ 307). The inherent power of clean metallic surfaces to repel scale, is gone directly the nucleus of a scale is formed. The water now acts on this in forming scale instead of on the hot plate, and this scale intervening is more effectually played upon than the plate was before, the heat being less intense. The behaviour of the water now will, although over an intensely heated part, be the same as at a part much less acted on by the heat, say, under the line of fuel with no film of scale intervening. Our heating surface now is one of scale instead of a clean metallic one. The process is now, that patches are being jerked off as the plate gets abnormally heated. The solids in the water will

immediately be drawn to that part, and will cover it as a flock of hungry birds would cover a patch strewn with food. The furnace sides of, say, $\frac{7}{16}$ plate have one half of their thickness heated to 1100° or nearly red-hot, the other half next the water to about 216°.

SCALE-DEPOSITING.

92. A severe contest is now being carried on by the separate forces acting upon each side of the plate (owing to the unequal expansion at these temperatures), the fuel endeavouring to heat the plate as hot as itself, the water to keep it down; and so long as the furnace is covered with clean water, the water will keep its ground, notwithstanding its tendency to fly off a hot surface, owing to the forced circulation by the hot plate. But if the water be densely salt in the boiler, say from $\frac{3}{33}$ upwards, the heated plate will arrest the salt in the water, through the medium of its salt-fluxed surface (§ 104), and as with a clap of the hand the salt is deposited on that part of the plate *en masse*. The water cooling it now is out of the question. The plate at once gets red-hot, and is bulged inwards by the pressure. The furnace now having lost that great natural maintenance of its strength due to the perfect circle, collapses.

Furnace Collapse.

93. (§§ 307-9.) At this crisis the collapsing will tend to break or shake the lump of salt off; it falls to the bottom of the boiler. If this happens the plate will again get cool; the water is now much less dense, owing to the salt extracted. This collected salt, however, will be gradually melting and resuming its place in the water, ready for a repetition of the process. The furnace now drawn inwards (see dotted lines, Fig. 34) to such a degree will bring an immense strain on the other parts, and seams of rivets, rupture, or explosion may take place along the line of seams.

94. In well-constructed good circulating boilers this crisis is not so much to be looked for, as the excess of common scale conduces to the salt deposit; but no boiler is safe with dense water.

95. It may safely be put down that collapse of furnaces is always due to deposit. Shortness of water is often blamed, but it is impossible the water could get so far down as, say, from four to seven feet, and the combustion-chamber top remain intact. The furnaces in tubed boilers are the last to be affected by shortness of water, and the first to be affected by density. This is the sort of accident (if such it can be called) that not infrequently happens to badly constructed, insufficient, or neglected boilers, known in North-country phraseology as "Stiff for steam jobs," and "priming jobs."

Virtues of Good Construction.

96. When circulation is an inherent property of a boiler, that must depend on the construction inside and outside, conjoined with its position in relation to its surroundings for draughts. Well-constructed boilers have good circulation, and good circulation prevents deposit of scale. Well-constructed boilers have good draught, and good draught prevents deposit of soot; good circulating boilers require no sediment catchers; sediment catchers impede the circulation which makes matters worse. (Continued at § 127.)

Salinometer.

97. The mechanical working of boilers by dead reckoning means the amount of water firstly admitted into a clean boiler, with the ratio of solid matter it holds in solution. With sea water the assumed ratio will be as shown at §§ 62–69, or, if thirty-three bucketfuls fill our boiler, one bucketful will be solid matter. We now keep tally of all additions to keep up the water-level, which can be ascertained and put down as so much per day. At any time this amount in gallons divided by thirty-three will give the number of gallons of solids in the boiler if no blowing out have taken place. If so, the duration of the cock being open and the pressure of steam would represent the difference between the result by reckoning, and the register of salinometer, supposing that instrument to be correct (which is not to be depended on). A

salinometer should always be proved before being used. Thus :—

98. Heat fresh or condensed water to boiling, wet the instrument in it all over, then, after allowing it to settle, mark the water-level. Then heat sea water to boiling, inserting and marking the instrument the same as before. Take the distance now between the two marks obtained; the first or fresh-water mark will be 0, the second or sea-water mark will be 1 or $\frac{1}{33}$; the first divide of the compasses will be $\frac{2}{33}$, the second $\frac{3}{33}$, and so on. In the event of a salinometer being broken or lost, the thermometer is no properly reliable test, owing to the fact that the ordinary variations of the atmosphere cause a much greater allowance to be made than there is margin of index available altogether (§ 221, P, T). Take a vial with a pencil through the cork to the bottom, having lead wire cuttings put in sufficient for sinking; or cut a piece of wood to the shape of a salinometer, twist lead wire round the under part sufficient for sinking, and mark either of these as here described for testing the salinometer, a good substitute for which you will now have got. Then :—

99. If our boiler is, say, 12 feet diameter, depth of water 9 feet, we lose by steam for donkeys, glands, and other quarters whence the water is not returned to the boiler, 3 inches of glass per day. Now 9 feet equals 108 inches, and divided by 3 equals 36 days; deduct one-third for convexity of boiler, tube and furnace space, and we have 24 days out of the original water, which is now gone, leaving its solids

in the boiler. What we have drawn from time to time—that is 3 inches of glass per day—has now amounted to another fill of the boiler, with its quota of solids. Our water therefore will be $\frac{2}{33}$, other 12 days' steaming will at the same rate make it $2\frac{1}{2}$-33rds, and so on, if no blowing have taken place. The density of the water ought not to exceed $\frac{3}{33}$, as at this temperature and strength of brine, sea water is no longer *sea water*. The heat now necessary to boil it is sufficient to bring about deposit of salt as described (§ 92), which means ruin and possible accident.

DENSITY.

100. Again, although the second deposit will take place as stated, it may also be superinduced at any point of density during the "interregnum," under conditions relating to scale as described (§§ 307–9). Therefore it must be a very mad game indeed to ignore the presence of salt in the water, and repudiate its destructive powers, while all the while it is hanging over our heads as a Damocles' sword.

101. The practice of freshening boilers by blowing out the water is herein proved to be a bad one, and in ordinary circumstances not required; but often circumstances of feed water being lost and steam wasted to an unusual degree, will of course bring about a corresponding density. Therefore it must be plain that "Don't blow a boiler," the creed of one sect of engineers, means a cry of "No wolf, no wolf," because for some time the wolf had not come out,

but with the wolf on their back there may be no cry at all.

102. Another sect of engineers believe in having firstly a scale over all the internal parts of the boiler, and if this cannot be had in the ordinary way the practice of

Artificial Scaling

is resorted to, the *modus operandi* being an imitation of the reality—that of opening the manhole doors and putting in lime, then hardening it on by heating the surfaces. Surely, in all reason, time and perseverance on the part of the lime contained in the water itself will accomplish this much-to-be-avoided result, without *raising our own hand against the boiler*. This practice, still prevalent in the North of England, is ostensibly to protect the boiler from rust, injury, and leaking. If the injury to be guarded against be that of "pitting," or rust, it seems very odd to take the "pitting" of one particular boiler for granted, and to administer as an antidote for one case what is bane in another, and a much greater evil.

103. If a boiler leaks, why should it not be repaired? It is well known that high-pressure boilers have a tendency to leak, owing to the expanding strain above the water-line (§ 212), and to have a boiler absolutely tight at high pressures is not easily attained; but it must be a curious leak that cannot be stopped. One writer on this subject remarks that it is a *peculiarity* with some new boilers that

they will not take on a scale naturally, and recommends artificial scaling. Boilers with this *peculiarity* are just what is wanted, and the pity is, if they be the exception. We have already said that a well-designed boiler meant a good circulating one, and that good circulation was the natural and best preventive of scale, or of "pitting." Therefore a boiler having this so-called peculiarity of good circulation would be spoiled by artificial scaling.

Fluxing Surfaces.

104. For instance, if we desire to join two metals together, say of iron and brass, by solder, we must first tin the faces to be joined. With resin as a flux we can tin the brass easily, but to tin the iron with the same flux is a long job; but after getting it tinned over, it will take on as much solder as you like. Having both metals tinned, the joining is now not of separate metals, but of one metal only. The tinned faces of each are united together, and if we had used tin instead of the brass, only the iron would require tinning. To unite metals together or substances to metals, or cover one metal with an amalgam of another, acids and other fluxes are used along with the heat. The boiler referred to would not take on an amalgam of scale naturally, therefore an amalgam was put on artificially as a means to an end. *Scale once acquired goes on apace.*

105. On other grounds the habit is inexplicable, as the parts that take on scale, viz. the heating sur-

faces, do not require it, and the parts where scale might be an advantage are just the parts where no scale will adhere, and will only be found there in very old *salt-seasoned jobs*, viz. under furnaces and on shell seams. The good old maxim, that "prevention is better than cure," is not easily upset.

This need not be confounded with the practice as a partial antidote against *rust* (§ 269). When a boiler has been allowed to get into such a diseased state, it is something like choosing the least of many evils, because heavy rust is oxide of iron scale, which as a non-conductor of heat falls little short of sulphate of lime. It must be evident that this is dealing with the effect instead of the cause (§ 186).

Salt-seasoned.

106. This term is applied to boilers that have been badly incrusted, having been repeatedly chipped all over. In such cases the plates of the heating surfaces are eaten in and impregnated with saline matter half-way through, and when once plates get in this way seasoned, their tendency to take on scale is increased fourfold, the substance of the plate itself being too much akin to the solids in the water. "Like draws to like."

107. When boilers get in such a way as to be thus taken possession of by the solids in the water, it must be plain that artificial scaling ought to be avoided.

108. Similarly as iron gets seasoned with the salt,

so does it get seasoned with rust. A piece of iron plate, bar, or bolt once thoroughly rusted will be eaten in and impregnated as in the other case.

It matters little how much be removed off the surface, that piece of iron will be rusted over, while a similar piece not previously rusted will maintain a polished surface (see Corrosion, § 269).

Priming.

109. Priming is a demonstration inside of a boiler, in miniature, induced by circumstances governed by the same natural laws as that of the waterspout, whirlwind, or hurricane on the face of nature. There is no phenomenon in the history of the steam boiler but can be put down as nature at large repeating itself in miniature.

The boiler is not merely a mechanical model in sympathy with nature, but a reproduction, and in accord with its laws as being part and parcel thereof.

There is nothing to remark as to a boiler itself pure and simple, that being only an inert vessel of iron. The whole matter is that it is a vessel, and if water be admitted (especially if charged with solids) on the application of heat this vessel becomes an institution in nature.

110. The ordinary causes of priming are better known than understood. The treatment of boilers is now so much a matter of routine that it is only extraordinary circumstances that bring about any-

thing very remarkable. Our *familiarity* with the general causes of priming is not to be confounded with our *ignorance* of the causes. Our individual experiences of the sensations of yawning, sneezing, and coughing, and the sudden strains brought thereby upon the animal frame, so often with ill consequences, would teach us that similar sensations in a steam boiler might lead to rupture and explosion (§ 332). So in the laboratory, so with the kettle on the hearth, so in the combustion of the fuel—draught and upward passage of the smoke.

So also from the same natural causes do the same inequalities of pressure, in conjunction with heat (in or out of the presence of water), abound in the boiler as in the atmosphere.

111. We are told that water finds its level, and that pressures equalise. But we must bear in mind that water seeking its level and pressures seeking to equalise are each of themselves forces, and any conditions calculated to put the boiler relations off the balance puts these forces into operation, and the water in seeking its level is thus kept in motion. So with pressures, from vortex to periphery, and from periphery to vortex.

112. We read that in some parts of the ocean storms have their regular paths, sixteen of which are in the Bay of Bengal alone, other parts with a more equal distribution of land and water are exempt, as also districts isolated from any land or currents underneath. We find the same phenomena in the boiler from a preponderance of gear, or space in one

direction, inside, or from unequal heating, sudden opening or shutting of dampers, stop-valves, fire, or tube doors.

113. Take an iceberg for example. It appears a great mass borne along by wind or currents, and apparently having no inherent force, and is as harmless as it appears in respect of that. But we observe a small turret at a far corner breaks off and falls into the sea. This has the effect of putting the whole mass off the balance. With a great crunching noise it tumbles, and keeps tumbling in all directions, creating a vacuum on the rising side, and beating the sea into huge square crested waves for many miles. And this demonstration, or expenditure of forces, is continued until the mass at last finds an equilibrium. Things just as comparatively trifling are known to start a boiler off priming.

114. Our planetary system could be also drawn upon to further illustrate the phenomena of priming. We may just as well take a lesson from our own physical constitutions by admitting into our stomach something not in affinity with something already there. We feel sick, our stomach seeks to void the irreconcilable part, in other words, we vomit it up. This is exactly what the boiler does, and we call it priming.

115. It is not uncommon for a boiler with clean water to eject every particle of mud from the bottom like a pellet by the waste steam pipe direct, and not blowing off at the time; at other times this discharge would be directed towards the stop-valve pipe

for the cylinders. Sometimes the water between bottom rows of tubes and furnaces is ejected.

116. A dangerous form is when the water and steam mix, showing no water, or otherwise, in the gauge glass. A lump of scale being suddenly released off, say, a furnace or combustion chamber, is often a cause, also steam pipe leading out on one side, and changes of temperature.

117. We have a simple illustration by putting a pan on a common fire. When boiling, if the water be dirty, it will fly up or prime. This is as with a dirty boiler. We cure this by putting the pan on a cooler place gradually, similarly as we ease fires by damper or otherwise.

118. Again, if we take clean water and drop in an egg while it boils, a priming commotion will take place immediately, which will also immediately stop the boiling, owing to the colder body reducing the temperature.

If the egg rolls about or if it breaks, the inside mixing with the water, there will be temporary priming. This also we cure by lifting it up to boil slowly.

119. For the outside the crusted soot on a kettle or pan is a benefit in the case of boiling milk, gruel, or any other mixed or heavy fluid. We also observe the non-conducting power of this soot crust (§ 150). We can hold the kettle in our hands with impunity while it is boiling. Unequal patches of soot outside, or any scale inside, or unequal flame will each cause indications of priming.

120. Furious priming often arises from change of water. This seldom happens in land boilers, but is experienced by steamers, say, coming from sea and getting into a river. They will often prime furiously, however clean the water may be; the same will occur with vessels going from river to sea. This kind of priming often acts in gulps, like a person sneezing or coughing. In such cases keep light well spread fires, ease the doors, as also the engines, keep a low head of water, and the priming will cease as the two waters get incorporated.

121. The habit of raising steam in too short time for the water underneath the furnaces to be heated often causes bad priming, until the temperatures be more assimilated (§ 333). The hydrokinator is used to circulate this cold water (§§ 265, 266). Locomotives often prime badly from, say, lime water at one end and mossy water at the other.

122. Where steam space is confined the intermittent drawing off or pumping of the steam, owing to the motion of the engines, is very effectually prevented by an intake pipe (Fig. 8) closed at end *e* and perforated or "slitted" on top side (making plenty of allowance for holes filling up) along the top of steam space, as in sketch. This has been long in use for marine boilers. In cases where the priming takes a whirling motion, the plan, as at B, is adopted as a preventive, viz. two iron angle flanges put in lengthwise.

123. Heavy scale on tubes stopping the circulation is another cause of priming. In this case the tubes

should be sliced at any rate vertically. If circulation is in this respect bad, a row of tubes might with benefit be removed to promote vertical circulation

Fig. 8.

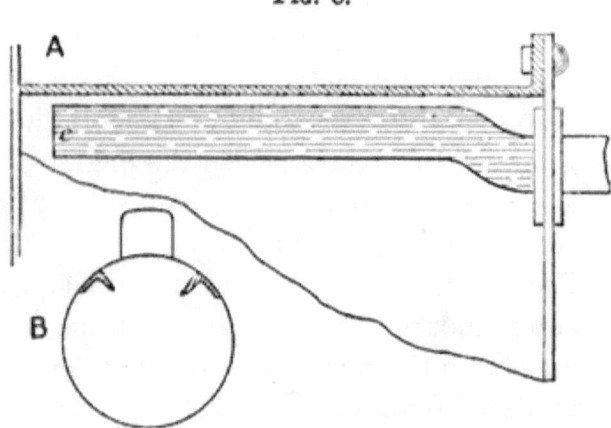

(Fig. 10, H, G, § 272). Everything outside or inside a boiler should have an equal distribution of surface, including the equal sweeping of tubes. Chemical combinations of the solids in water produce priming under favourable conditions.

Slight priming will agitate the mud in the boiler bottom, which, when mixed with the water, produces heavy priming. A white squall is a good illustration of priming.

124. In addition to the foregoing causes for priming, working singly, two or more boilers working connected sometimes give trouble, owing to one setting off the others, from such causes as one generating more steam than the others, or stop-valves not being equally open, unequal draught, one having

too much water, thus contracting the steam space; the pumping influence of the engines added to this is cause enough. One having the water too low will, with heavy fires and long-flaming coal, have the water forced off the plate by the intense heat, owing to insufficient head of water (see Lifting Water, § 232).

125. To prevent priming generally, avoid all the causes here referred to, including any others of a kindred nature. Have a regular methodical way of working. Use clean water, regular firing, steady steaming, looking out for what follows, so as to avoid all sudden stoppages, sudden opening of valves, doors, and dampers. Avoid suddenness of anything, and always promote cleanliness inside the boiler.

126. To stop priming ease all fire-doors a little, check the draught by also closing the damper a little, slow the engines and open drain cocks, slowing the engines gradually until priming ceases. It must be obvious that slowing the engines will stop it, but endeavour also to prevent blowing off. Never use grease or tallow to prevent priming (§ 232).

The causes and cures for boiler priming are, as with the doctor and the human subject, deduced from the symptoms. The engineer, on examination of his boiler inside, if after priming, will find much in the way of symptoms, especially about the shell crowns, which, with other observations, may point very nearly to the seat of priming.

CIRCULATION.

127. The grand circulation constantly going on in the ocean and in the atmosphere, mainly due to the expansion of water and air by heat, we have in miniature in a steam boiler. Good circulation of the water constitutes all the difference between a good and bad boiler. As the circulation must naturally be in vertical lines from and to the heating surfaces, any complication of tubes and gear interrupts the ascending and returning messengers of heat and steam; the ascending can thus only prevail by the intensity of the ebullition. The construction of many boilers is such that there is no go and return path for the circulating particles. Indeed, structurally considered, there is no plan of circulation at all. At sea the circulation is always best promoted in a rolling ship, the conditions being still more complete in a revolving boiler with fixed fires, which, if otherwise practicable, would fulfil several natural conditions against priming and deposit, however dirty the water.

With bad circulation we have different pressures in the boiler at the same time. These pent-up forces lurking here and there, especially in the region of the tube spaces, are also incentives to priming. Tubes too closely set produce chronic bad circulation, and consequently priming.

128. Neglect is the common promoter of bad circulation—tube spaces $1\frac{1}{2}$ inch when new, being allowed to scale up until a space-lamp $\frac{3}{4}$ inch thick cannot

be got between them, as at Fig. 10, C. Now, we know that 60 to 75 per cent. of the heat produced must pass through these spaces by the water to the surface, and that as we allow scale to increase so do these circulating spaces decrease and choke (§§ 150–152). Therefore at the top row of tubes the circulation is throttled, and it is only the increasing pressure in the water space over that of the tubes that gives what circulation there is.

129. There are places over the face of the ocean where circulation is flat and sluggish, and in several parts of our home seas, such as Ailsa Craig and others, where there are no perceptible tides. So also is it in a boiler naturally, owing to the influence of surface distribution over the circulating currents, generally under the furnaces, about the water-line of shell and other places. If we be so ill-advised as to insert sediment catchers in such eddies this will have the effect of further reducing the circulation, and make bad worse.

130. The action of circulation can be very well illustrated with small vessels of different construction to represent the various descriptions of boilers, over a common fire with clean water in which is put a little bran, which will faithfully demonstrate the circulating properties of each.

131. The circulation of the water in a boiler determines the various functions of which the individual boiler is capable—distributing the heat, preventing deposit of solids, carrying off all humours, causing pitting, &c.

These are just exactly the functions performed by the circulation of the blood in our own bodies, as well as those of other animals, and are also faithfully reproduced in other departments of animated nature.

132. Three furnaces in one boiler is a better adaptation for circulation than two. Of the present boilers in practice the upright, or donkey boiler, is the best circulating of all (§ 298).

It is in the badly circulating boilers, or badly circulating parts of a boiler, that we have to look for that insidious boiler disease,—

133. PITTING,

which is brought about by patches of moss, leaves, bark, or other organic matter containing tannic and other acids, getting into the boiler with the water, and afterwards adhering. By the action of the acids, assisted by galvanic and other influences that prevail in places where the circulation is sluggish, these patches get resolved into forms like barnacles, and adhere to the boiler shell like these latter to a rock. We might almost assume for these a low form of life, such as the barnacles. These "parasites" are generally found in a quiet and not too hot place, just such a place as Scott would place his sediment catchers in. They directly attack the plate through the medium of the acids they contain, and very soon form a hole or pit by decomposing the iron into a brown viscid semi-fluid.

A. The chief engineer of the s.s. "T——" procured a sack of oak sawdust as a scale preventive for his boilers. Two bucketfuls, and water added, were put into the boiler by the condenser when under weigh. Four weeks afterwards the boilers were opened out, when a great number of these barnacles were found busy at work from the high-water mark downwards. This oak sawdust, like similar substances in water, had got into knots or patches, and floated about (with something like the instinct of a fish to spawn) until it found a quiet place. Several of the largest of these were $1\tfrac{3}{8}$ inches across when removed. The holes they were making were $\tfrac{1}{2}$ inch to over $\tfrac{5}{8}$ inch diameter, perfectly round, as if bored with a drill, and $\tfrac{1}{4}$ inch to $\tfrac{1}{2}$ inch deep, and were about half filled with this acid spoken of, which must have had its germ from the tannic acid of the wood, but had increased its substance from its own operation. These ulcers were cleaned out afterwards, and swarfed with a drill until the pure iron was reached, then filled up with Archangel tar and cement. Another kind of these parasites were at work in a straight line, forming figures similar to a chain. They were very much smaller, and not so destructive. That "chief" did not want any more sawdust for his boilers. Seeking in oak sawdust a preventive for scale, he was counteracting one evil at the expense of risking a greater.

B. An egg-end boiler in the county of F—— was ruined in two years in this manner. It was only used about seven months in the year, and was fed from a moss ditch. There was no scum-cock.

The attendant simply ran the water off when cold. This thick mossy scum of course followed the water down, covering every nook and corner in the most complete manner, the bottom having a double dose. This, then, remained five or six months undisturbed. The attendant, in his ignorance, made it a point to wash out the boiler when the working season began, whereas he should have washed it out when the season ended, which is just as the man's own wife would do after a day's washing, have the boiler washed out with *soda* to keep it from rusting. If he had just consulted his wife he might have saved the boiler, which was perforated like a honey-comb with many thousand of these ulcers, or pits, and was soon afterwards condemned.

134. We can have a very good illustration of the pitting process with a wash-hand basin, say, connected with an engine room, where soap has been used to clean greasy hands repeatedly, and this water afterwards allowed to remain undisturbed. By and by the basin leaks. Pouring off the water we will have a thick residue at the bottom, and find numerous spots around the bottom edge from the well-hardened scab, or outside skin of the ulcer, eating right through the iron, down to the more primitive stages. On this basin being well washed it will be found riddled with holes, and otherwise marked with some brown spots, indicating the earlier stages of the process.

135. These ulcers will be found in bad circulating quiet places in the following order (see Fig. 9):— Furnace bottoms, $d\ d$; undersides of tubes

Pitting. 77

(especially if welded seams are so situated), along the line of fire-bars, ee; then the shell, as at ff, ff; and under combustion chambers, gg; also front end plate within water-lines, hh. The chamber top is exempt, as also the upper parts of tubes, owing to the greater heat being unsuitable for these ulcers. The illustration of the wash-basin can be tested at any time, but that of the oak sawdust had better not be inflicted upon any boiler. At any rate, both references prove important facts, from which we may deduce the following lessons:—

136. Never to use up soap-suds or washing water from hands or clothes, or water containing any soap

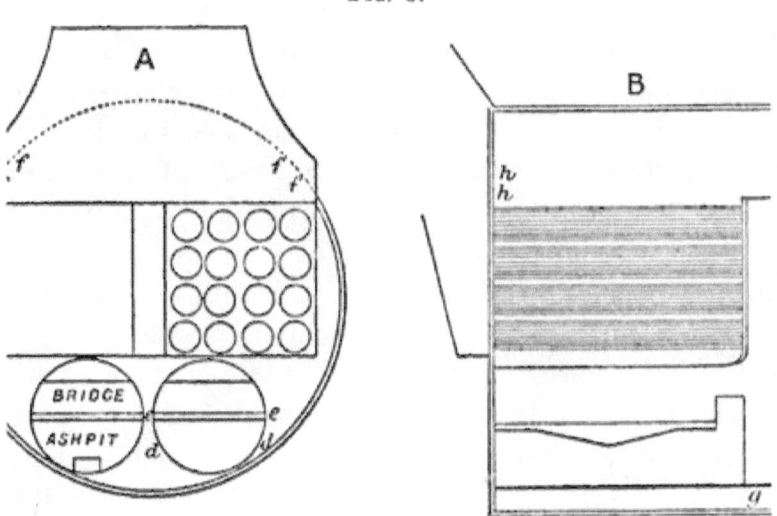

Fig. 9.

whatever, for boiler purposes. Be careful of what you put into a boiler of any substance whatsoever (see §§ 205-208 for reason). Use as little fatty matter, such as cylinder lubricants, as possible. Tannic acid means the acid of barks or wood.

Soft water always contains acids, producing pitting. No better preventive against pitting need be than soda (§ 185).

STIFF FOR STEAM.

137. Amongst the engineers of our ocean-going steamers, or common carriers, which includes the bulk of our mercantile marine, is a common phrase, viz. "Stiff for steam." Very often this phrase is affixed or prefixed in North Country phraseology as expressive of the surroundings of the job as well. " Stiff for steam " comprehends a great deal. Commercially, it is a question of first cost, secondly, a loss by extra tear and wear, a grief to all concerned, and a white elephant to the owners, who, to begin with, wanted a vessel of a certain carrying capacity and fixed draught to travel at a given speed on a fixed quantity of coal per hour. They receive tenders from boiler builders, who in their competition against each other get the job fixed at the lowest margin of profit, or none at all. The trial trip is run with, say, 5 per cent. to spare off contract time. In all cases everything is done in the engine room and stoke-hole to exceed the contract. But running the measured mile or ten miles is no criterion of the average voyaging speed covering a period of four or five years. The dictum of builders is so much fire-grate and so much heating

surface yields evaporative power sufficient for x horses' power on x tons of coal. This dictum is received by the owners as reasonable, and they admit it proved at trial trip, viz. she ran $12\frac{1}{4}$ knots, nearly a knot over contract.

138. Well, supposing the boilers are perfectly capable of this rate of steam generation when brand new (§ 159, C), four years after, what of it then? What of the heating surfaces, combustion chamber, tubes, and tube-plates?—tubes with $\frac{7}{16}$ of scale and $\frac{5}{16}$ soot average owing to hard firing (§ 150, F, G, to 154, B). How with tube-plates buckled and tube-ends leaking at such a rate as requires donkey to keep water in boiler, if auxiliary is not on? How with combustion chamber backs bunged up over the bridges with muck and ashes salted into stone from leaky tubes; total of fuel lost 70 per cent., and loss of speed equal to 190 per cent. of coal? The speed of vessel proved at first $12\frac{1}{4}$ knots, now she won't do 9.

139. The engineer, who possibly has only been in the boat one trip, and has not got a chance to be inside of the boiler, will answer, "Stiff for steam." On examining his boiler he finds he has got a *salted job*, tube-plates buckled, tube-ends leaking, heavily scaled tube-spaces, everything else in keeping. The vessel has to go next day; no use telling the owners to delay; he can only have his boiler shut up and content himself with perhaps writing a report (often harmless things), with a view of either having boilers properly scaled, or else leaving the ship. Previous to his joining she could only travel with *hard firing*, and

as he must not let down the speed he has to resort to hard firing too, and hard firing means no intermission from shovelling, poking, slicing, and pricking. Instead of clearing one fire per watch with 4 or 5 buckets of ashes, it means two or three with 17 to 24 buckets of ashes, or, in other words, *murdered fuel* knocked through the bars or up the chimney not half consumed, through the process, as if getting rid of the coal was the object aimed at.

140. "Stiff for steam" means low speed and low profits owing to high bills for coal, stores, and boiler repairs. There are few boilers which can under average circumstances generate more steam than the engines can devour. The boiler power to be the first to fail might be originally a prudent safeguard. It may prove so for the engines, but the reverse for the boilers themselves.

141. It is not enough that a job be run at, say $10\frac{1}{2}$ knots, with hard firing and all its attendant evils increasing, and its speed decreasing in a year down to 9 or 8, whereas $10\frac{3}{8}$, 10, or $10\frac{1}{4}$ knots could be run steadily with proper combustion economy and safety with a boiler very little larger. The last knot, and the last $\frac{1}{2}$ knot, $\frac{1}{4}$ knot, and last $\frac{1}{8}$ of a knot is all the difference in the world between stiff and easy for steam. Speaking roughly, this last $\frac{1}{8}$ of a knot costs nearly half the whole fuel consumed in a case of "stiff for steam," or about a ratio of 200 per cent. excess of complete economy (§ 150). Therefore the more a boiler is forced the more unfit it becomes for duty.

142. There is no secret in firing boilers known to one and not another. The blowing-off pressure, the mean pressure of steam carried, and the pressure the job steams best and most economically at, are three separate points.

143. Speaking generally, boiler builders do not assume that their boilers would be so scaled as to place their heating surfaces *hors de combat*. Keeping the boilers clean is no part of their bargain, and in the first instance would mean a more expensive boiler, and this is where the shoe pinches, the owners counting at first cost. A larger area of fire-grate would allow of more air passing by the fire-doors over top of fire, uniting with the gases in the combustion chamber, re-supplying them with that quantum destroyed in the passage through the fire.

144. There is certainly no saving of space in having small boilers (§ 26). The larger boiler with easy firing burns so much less coal, as by slower and better combustion the fuel is not belched up the chimney in such black smoke, nor is there thrown over the side as ashes what otherwise would be "burnable" coal. This being the case, we can dispense with bunker space to pair off with the increase of space for boiler, thus saving the fuel. Small stiff boilers require large bunkers. Stiff boilers are always bringing grief.

Large easy boilers will run ten days, without cleaning tubes, easier than a small stiff boiler will run three days.

Soot.

145. Another inveterate non-conductor of heat, very prevalent in boilers, especially tubed boilers, is soot, and its property of non-conduction is greater even than that of scale. Much less attention has been given to this loss than the exigency demands. Scale is looked after because its presence in quantity is dangerous; thus brought to the front its non-conducting property is always paraded before us. Soot again, because its presence is not dangerous, is not even popular as a subject, nor its prevention considered as a factor of economy. And true it is, that scientific writers underrate or ignore the subject, notwithstanding that it is the greatest non-conductor of heat affecting us.

146. If we ask any sea-going engineer or fireman how it would benefit them having to clean no tubes, or not having to fire boilers with dirty tubes, they will reply that they do not grudge the trouble of cleaning tubes, notwithstanding the mess, as they are more than repaid by the easier firing afterwards. But cleaning tubes (owing to heating surfaces having to be exposed) means low steam and a ship going $1\frac{1}{2}$ hours about half speed. The engineer must endeavour to avoid this until it is forced upon him, and certainly any engineer out of a "stiff job" (and most cargo boats are stiff enough) would give a hearty "hear! hear!" to any denunciation of soot, which would be echoed by the firemen.

147. That soot is a non-conductor is easily proved.

Take a tea-kettle that is besooted up the sides and that is boiling on the fire. You can clasp it, holding it up with your hands for an unlimited time, the water inside at 212°, with the outside half of that heat. You cannot do that with a clean surface.

148. Again, take an earthenware pot clean, as being of the nature of boiler scale, and this we cannot hold in our hands. Again, the presence of soot influences in great measure the draught by the ragged or fringed surfaces intercepting the smoke, independent of the reduction of draught area.

Soot and Scale Effects.

149. For instance, a furnace stalk of sheet iron 14 inches diameter has a coating of soot, say $\frac{1}{8}$ inch thick; available diameter now 12 inches. A pipe 10 inches diameter clean will be as efficient as the 14-inch one with soot. Or a clean stove-pipe 4 inches diameter will be as good as a dirty 5-inch one. A common chimney when dirty has less draught, in proportion to the respective differences of area. Thus, a clean surface and small area has better draught than a larger area with dirty surface, which is a well-known domestic fact; in other words, there is no comparison between a dirty and a clean chimney. Fig. 10 is an illustration of the effects of scale K and soot L, separate or combined. a, b, c, d are two boiler tubes; X c, d represent the tubes clean; H a, b represent them salted and sooted to the last degree.

84   Steam Boilers.

150.   Fig. 10.

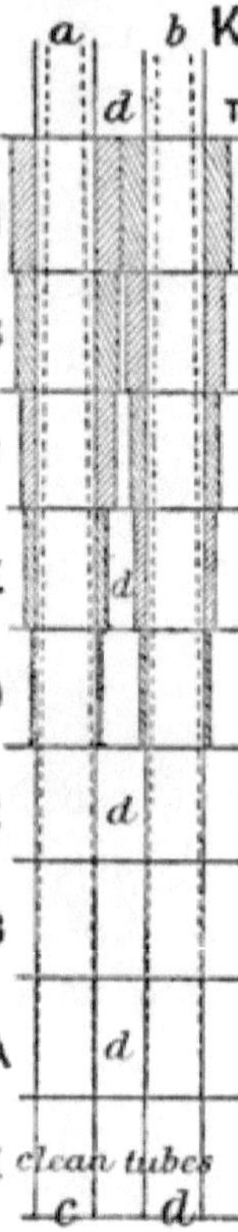

	a	b	K SCALE LOSS %				L SOOT LOSS %				GRAND TOTAL
		d	THICK	NONCON-DUCTION	NONCIR-CULATION	TOTAL	THICK	NONCON-DUCTION	STOPPAGE OF DRAUGHT	TOTAL	
H											
G			½"	100	40	140	½"	140	20	160	300
F			¼"	50	20	70	½"	140	20	160	230
E		d	⅛"	24	10	34	½"	140	20	160	194
D			1/16"	10	7	17	⅜"	80	10	90	107
C		d	1/32"	4	3	7	¼"	50	6	56	63
B							⅛"	20	3	23	23
A		d					1/16"	10	1	11	11
X clean tubes											
c	d										

Line A represents 1/16, B ⅛, C ¼ inch of soot.
 „ D   „   scale 1/16 inch thick and soot ⅜.
 „ G   „   „   ½ inch thick and soot ½.
 „ H   „   the circulation through tube spaces d d d stopped with scale and bunged up with soot, at which extraordinary state of constipation we will melt the iron of the heating surfaces sooner than heat the water.

Soot and Scale Effects.

	Loss Per cent.		Per cent.			Loss Per cent.
Scale E of $\frac{1}{8}$ = 34	}	combined	90	{ Scale as about C	8
Soot C ,, $\frac{1}{4}$ = 56				Soot ,, A	11
						19

This 19 per cent. being good average work.

At H the only circulation is between boxes of tubes.

151. If we take a clean boiler, say brand-new, and put clean tubes as 100 per cent. economy, and putting soot of $\frac{1}{2}$ inch from fringe to plate as $\frac{1}{4}$ inch, this means 45 per cent. loss on account of soot. And with $3\frac{1}{2}$-inch tubes the difference of area for draught will be $12\frac{1}{2}$ to 9, or, as 100 and 72 respectively, or 28 per cent. for decreased ventilation or tube draught. Then, if we put decreased tube draught (as affecting the consumption) as $\frac{1}{3}$ that of soot, and loss by non-circulation through water spaces (owing to being choked with scale) as $\frac{2}{3}$ that of soot; and put non-conduction of heat, through interposition of scale, as 127 per cent., assuming other heating surfaces as equal,

152.

Then we have non-conduction of heat by soot = loss	45	
Decreased tube draught one-third of 28 per cent. = loss	9·34	
		54·34
Interposition of scale on tubes three-eighths in thickness = loss	127	
Non-circulation through water-spaces owing to being choked with scale = loss	18·66	
		145·66
		200 per cent.

Or three times the former consumption, as at § 150, which was 15 tons per day at 11 knots per hour, or equal to 45 tons per day at the same speed, which we cannot get. Moreover, $22\frac{1}{2}$ tons is the most we can knock through the fires in 24 hours with hard firing.

At the 15 tons consumption our ashes were 10 cwt. or $3\cdot 3$ per cent.; at $22\frac{1}{2}$ tons, 120 cwt. or 40 per cent. of the original consumption, or $13\cdot 33$ per cent. of the increased consumption. Deducting original ashes $3\cdot 3$ or 10 cwt. per day from the increase, leaves 36 per cent. to the account of "murdered fuel" (because the fuel is not consumed) thrown over the side, with a corresponding amount forced up the chimney. This is what becomes of the coal. Therefore this "murdered fuel" is a separate account consequent on hard firing, but, of course, induced by the "stiffness for steam," owing to the soot and scale. Then, for the remaining coal unaccounted for, that has gone to increase of temperature of funnel gases. The result of all this is now for

$22\frac{1}{2}$ tons .. .. 10 knots.
15 ,, ... 9 ,,
10 ,, ... 7 ,,

Compared with formerly

15 tons .. .. 11 knots.
10 ,, ... 10 ,,
$7\frac{1}{2}$,, ... 8 ,,

153. We will take a vessel burning, say 1000 tons of coal on her first voyage (a stiff job) for 10 knots. After three years the same voyage requires 1750 tons

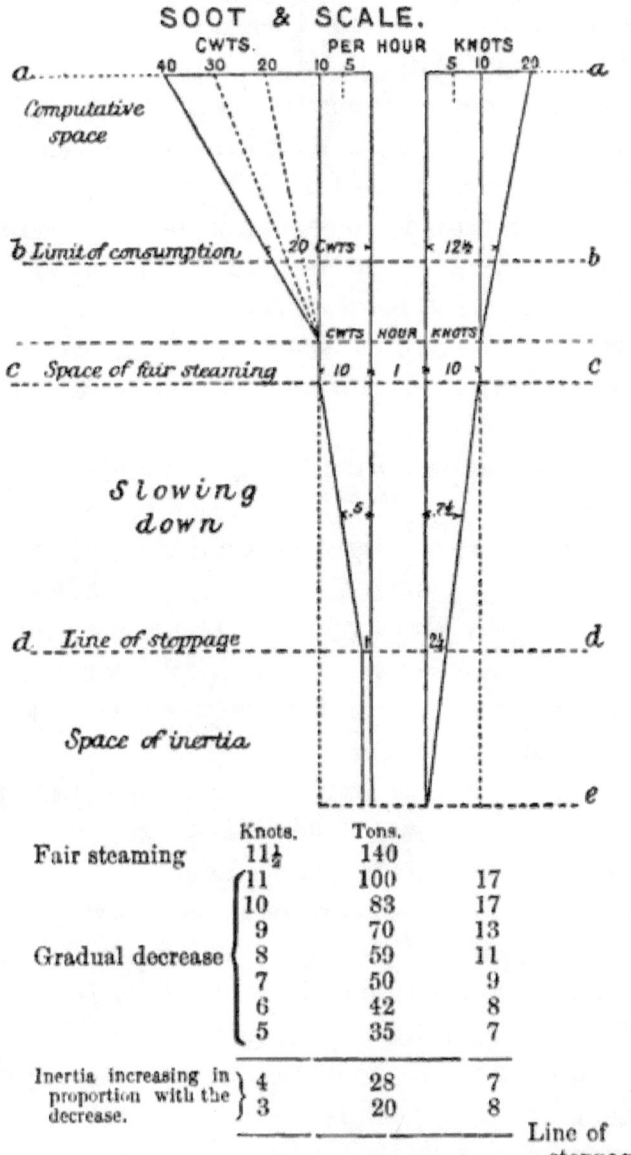

Fig. 11.

As the ratio of friction increases in a similar ratio, therefore at 4 knots an hour more is expended on friction.

for 7¼ knots. Thus, first voyage, 1000 tons, at 10 knots; owing to scale and leakage the last voyage, 1750 tons at 7¼ knots.

The difference between the last and first, 2¾ knots; a speed of 2¾ knots per hour being equivalent to the increased consumption for the last voyage.

Now, our rate of consumption latterly, compared with the original speed, is 1000 for the voyage at 10 knots. This 1750 is *all we can put through the fires;* therefore, taking this same rate and adding the equivalents A B gives the ratio of 5 to 1, although it is not realised.

154. The figure to the left A is a diagram of fuel consumption.

The figure to the right B is a diagram of speed.

The space between C is a diagram of 1 hour of time. Thus A = cwt. of fuel. C, per hour for units of work B, or A = cwt. per hour, for knots per hour. The top line A from 1 to 40 represents 40 cwt. equal to 40 inches long.

The line A b represents the greatest consumption as 20 cwt. per hour. The line A c represents consumption for ordinary fair steaming, 10 cwt. per hour. The line A d represents line of stoppage or inertia as 1 cwt. The space d to e represents that a continual consumption of 1 cwt. per hour is expended on friction or inertia = $\frac{1}{10}$ of fair consumption. The vertical line 1, 1, from line of stoppage also represents 40 cwt. The diagonal line 10 to 40 represents the ratio of increased consumption for increased speed. The slanting line downwards from fair steaming to line of stoppage represents decreasing consumption down to 1 cwt. per hour, for $2\frac{3}{4}$ knots speed B d from this point d to c.

The top line B, 1 to 20, is just half of the corresponding line A; this is 20 knots or 20 inches long. The line B b represents the greatest speed = $12\frac{1}{2}$ knots. The space B c represents ordinary fair steaming speed, 10 knots. B d represents the line of stoppage as at A. The space d to e represents that useful effect equal to $\frac{1}{10}$ of consumption is expended in friction. The vertical line 1, 1 from line of stoppage also represents 20 knots. The diagonal line 10 to 40 represents the ratio of increased speed.

The slanting line downwards from fair steaming to line of stoppage represents decreasing speed as far as $2\frac{3}{4}$ knots per hour, then the engine stops. C represents simply a period of time as one hour. Again, taking the two diagrams A and B side by side and comparing them, we find they are only equal at 10 or space of fair steaming C. The preponderance in favour of A, from c upwards, shows the unequal ratio of consumption for increased speed against B.

155. Again, from line c downwards to d we have another ratio, this time the preponderance in favour of B. At C or space of fair steaming, speed and consumption are balanced. As we slow down the ratio of consumption decreases, as at D or line of stoppage; one cwt. yields $2\frac{3}{4}$ knots in one hour. § 155 is a parallel illustration in figures. Fig. 10 is a tabular diagram representing loss by soot and

scale. § 151-2 is a parallel case with Fig. 10, illustrating ratios of speed and consumption.

156.

1st voyage			100 tons at			10 knots.
2nd ,,	..	..	110 ,,	..	..	10 ,,
3rd ,,	..	..	112 ,,	..	..	$9\frac{3}{4}$,,
4th ,,	..	..	114 ,,	..	..	$9\frac{1}{2}$,,
5th ,,	..	..	117 ,,	..	..	$9\frac{1}{4}$,,
6th ,,	..	..	120 ,,	..	..	9 ,,
7th ,,	..	..	124 ,,	..	..	$8\frac{3}{4}$,,
8th ,,	..	..	127 ,,	..	..	$8\frac{1}{2}$,,
9th ,,	..	..	131 ,,	..	..	$8\frac{1}{4}$,,
10th ,,	..	..	137 ,,	..	..	8 ,,
11th ,,	..	..	145 ,,	..	..	$7\frac{1}{2}$,,
12th ,,	..	..	153 ,,	..	.	$7\frac{1}{4}$,,
13th ,,	..	..	164 ,,	..	..	7 ,,

A vessel going 11 knots with two boilers will go 7 knots with one boiler and half fuel.

If heat could be gathered up with the same facility as that with which at every point in engine-room and stokehole it is scattered, one great desideratum would be attained.

157. These computations (approximate) are not overdrawn. The following illustration, less speculative, that of a common stove, will show the practical effect of tubes as a heating surface (Fig. 12). S, stove; e, top or lid part gets hot first, then a receives the flame in half-flat contact. The part of pipe about d next the stove is heated by conduction from the stove itself. From a to b there is little development of heat, but at c we have the effect of the concentrated flame and heat in the same angle of contact as at a depositing a large percentage of heat. A reactionary effect is produced at b, being the point of spring,

therefore the economical effect of the knee in the pipe is doubled. The straight part between *a* and *b* illustrates the value of a boiler tube as a heating surface (Fig. 28, C). The heat is being carried off as

Fig. 12.

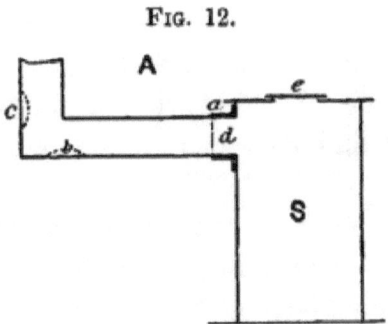

by an insulated current, enveloped by a shroud of smoke and escapes; thus, by the length of tube and smallness of diameter is this waste of heat promoted. However correct, then, may be the principle of having the body of water in a boiler intersected by tubes of thin material, the practice is subject to serious drawbacks.

Feed.

158. The feed water ought to be injected immediately over the tubes in a pipe extending nearly across the boiler, closed at ends and perforated to throw the water into the water space, and in no case should the pipe, if copper, be allowed support on any tubes or other gear of iron in the boiler without a piece of lead being put between, otherwise, owing to the galvanic action, the malleable iron would soon be eaten away.

The water itself must not be allowed, when entering the boiler, to impinge upon any of the internal parts or iron whatever. Main engine and donkey feeds should both lead into this pipe. The action of the feed should as much as possible circulate the water about the tubes, &c., crosswise; it ought not to have an upward tendency into the steam space, because of saturating or charging the steam with water, which, when it find its way into the cylinders, is not a nice job.

159. With surface condensing engines having feed pumps and checks in good order the feed water gives little trouble beyond the trying of gauge cocks, and adding to the boiler from the " supplementary " for loss by glands and other sources of waste. In high-pressure engines the firing, feeding, and trying of engines comes as one, two, three; one, two, three. But with more than one boiler off the one feed, or with a mixed feed, inattention may involve the greatest danger. The phenomenon in this case is similar to that when showing false water (§ 230), and the consequences identical.

160. Suppose a case of feeding two or more boilers at the same time with the same pumps, and say the water is at half glass in both boilers; the attendant is absent fifteen minutes; on his return the water shows *out of both glasses*. Now, if the attendant is not " up to " the cause of this phenomenon, or in other words, not master of the situation, it is a *crisis* with one or other of the boilers. If he be inexperienced he will get confused and alarmed, and possibly accelerate the result.

Here is a case in point. The attendant returning after a short absence, manipulated the gauge cocks, and found them all apparently blowing steam. He sent for the chief engineer, the firemen meanwhile drawing fires in both boilers. The chief arriving on the scene, tried all the gauge cocks properly, and found that in one boiler all cocks were blowing steam, and the other all water. All feed was now shut off this boiler on to the other, fires set away in the former, and engines kept going. By and by an indication is shown at top of one glass, and shortly the other shows water at the bottom—fires now all set away. In twenty minutes more the water is at half glass in both boilers. There was no accident resulting from this, or accident to cause it. The explanation is, that the attendant had over-done the checks when reversing the feed; also a fire had been cleaned and bridge repaired in one boiler, which, of course, was generating less steam than the other, and receiving the full feed, and, following the law of all liquids under pressure, it took to the opening having the least opposing force.

Now there was nothing very culpably neglectful or careless in what the attendant had done, his experience considered. But the consequences of his misadventure might have been bad enough. It is not a *very* good maxim, " All's well that ends well." The culpability, however, was enough to entail very serious consequences, and the incident ought altogether to be a life's lesson to him. A boiler will not be neglected with impunity. The first look-out of an engineer or boiler attendant is the water in the glass and that the cocks are free. Cases such as that referred to are of too frequent occurrence. With all kinds of boilers or engines the feed is all-important.

BLOW-OFF COCKS.

161. All boilers are supplied with blow-off cocks, top and bottom, the top one being termed the scum-cock, the lower one the bottom blow-off; the intention of the top one being to scum or lead off any organic or light matter floating about the surface of the water. By judicious use of this cock we can get rid of the bulk of this matter, and if using mossy or surface water (§ 183), this cock is of good service in ridding us of the noxious acids contained in such matter likely to produce pitting. But it must be a rule that this cock be only used when absolutely necessary for scumming. If in river work using much water, two inches or less per day of the gauge glass may be blown out. With vessels using sea water, half an inch every two days is enough to carry away any light muck, which is chiefly the fatty matter of the cylinder lubricants. This practice also keeps the cock free, but for partially changing the water, termed "freshening," this cock must on no account be used.

The changing of water must be done by the bottom cock, except in cases of acidity (§ 183).

Too much use of either of these cocks subverts the purpose for which they are intended. One important matter is, that they be kept tight as well as free, that is, that they do not leak. They should be cleaned and ground frequently; an occasional "crack! crack!" from these pipes indicates they are leaking.

Blowing Practices.

162. There is much conflict of practice in using these cocks owing to conflict of teaching. One school says, "Blow not a boiler"; another school says, "Blow away while you can." Another school works on with a leaf out of two books—a reconciliation of the irreconcilable (§§ 72-5).

163. Now by neither is a word said about the feed-pump, or valves, or leakage of glands, or pipes on deck. What about the difference between ocean-going and coasting steamers? What about the length of distance run, and length of time a vessel has to lie at each end? And what about vessels that have no donkey boiler? What amount of fire-grate per horse-power and consumpt of coal per hour? Is your job stiff for steam? What amount of heating surface have you to every 100 gallons of water evaporated? or, with steam up, what percentage of the full pressure is carried? How much blowing-off of steam? How long under banked fires?

Then as to the water question:—

Blowing out Water.

164. When a boiler has been filled in part to say three-quarter glass, the raising of steam will expand

the water (§ 228) to full glass, or nearly. Before starting the engines, if using sea water, blow from bottom 3 inches of the glass. This will carry off the carbonate of lime, and whatever sulphate has been deposited (§ 81). When twenty-four hours under weigh blow one inch of the glass by the scum-cock to carry off any matter on the surface. If the water in the boiler has not been excited by stoppages or priming, blow another $1\frac{1}{2}$ from the bottom. If at sea, all going well, this cock need not be used again for eight days or more. If using fresh water, blow out the same amount, but using the scum-cock more and the bottom cock less in proportion to the nature of the water. As carbonate of lime is deposited before steam is up, and also the sulphate, the sooner they are out of the boiler the better, as in the bottom they are an incentive to priming, and if any commotion or priming takes place these solids, now as a white mud, will be largely resumed by the water, and after priming ceases these suspended solids will, when again depositing, put another stratum or coating all over the internal surfaces, this process being repeated each time the water gets excited. And if the vessel is brought to anchor, or to port, and boiler cooled down, this mud will be resumed by the water as it cools, to its full carrying capacity, and when fires are again set away this depositing process gives another coating. Therefore it is plain that this deposit of lime should be got rid of as soon as deposited.

Changing Water.

165. The sovereign way to change the water in a boiler is, first, to allow the fires to die out; when steam is reduced to about 25 lb. scum the water to bottom of glass, shut the scum-cock, and allow the boiler to cool for three days; then run it out into the bilge, and take off man-hole doors, both top and bottom, one of each, more if necessary. If three days cannot be allowed, try with two, or one and a half; as much as you can get.

166. For greater expedition, the next best way is to reduce the pressure to about 20 lb., blow the water from the scum-cock as far as it will take it, and start the donkey pumping water in, which prevents the scale drying; open the bottom cock and blow down to about four inches above the furnaces, or about the level of test-cock, and then shut off, leaving this in the boiler when the steam is off; open the gauge-glass cocks to allow air to get into the boiler to prevent a *vacuum* being formed (§ 267). Allow this water to remain at least 12 hours, if possible, or more; indeed, as much as you can; it can then be run into the bilge; keep the man-hole door on as long as you can afford. By doing so you keep the scale soft. Remove man-hole and bottom doors; if wanted to cool quickly, take all doors off. If, however, scaling is to be done, take off the bottom doors only; this will prevent evaporation; the scale will therefore be damp and easy of removal. There is an immense difference between a damp soft scale and

a hard baked one, which latter is caused by allowing the water to cool in the boiler without blowing it out, and of necessity having to blow it out forthwith, the heat following the water down and drying the scale hard on. To allow the internal heating surfaces of a boiler to dry quickly is just a trapping of the suspended salts in the water into hard intractable scale.

167. When blowing out water always in the first place open the sea cock full open, then the boiler cock slowly. If it makes much noise reduce the opening. The reaction in some boilers is great from the impact of the hot and cold water, as also from the sudden expansion of the pipes. The tubes and other parts of some boilers leak every time there is blowing done.

168. In every case where the water is blown out of a boiler it should also be well washed out with a donkey hose or other vigorous means (§ 173). If any pieces of scale be cracked and likely to fall off, have it all removed forthwith, whether scaling is to be done or not (§ 174). If scaling is to be done, and the crust soft, the task, according to § 166, is simple, although tedious with tube boilers (§§ 271-3).

But if hard it becomes a hard job, and many expedients (some of them very questionable) are resorted to, to facilitate the operation. The most vicious and destructive to the boiler is the *flogging method;* that is, a pair of fore hammers are applied inside and out at the same time, all over the plates, which cracks and shakes off the scale. Much as this is to be condemned, it is, however, in some cases the only way available, notably with Galloway tubes.

169. Another and less objectionable plan is to expand either the scale itself until it looses its hold, or the plate or tube until it bursts the scale.

170. The least objectionable for land or tubular boilers is, if the boiler is cold, to set away a good fire of sticks and let it die out. Half an hour afterwards repeat the fire. This allows the first heat to enter into the iron.

We shall then have to be content with the result, which often brings off a mass of scale.

171. For marine boilers a large choffir or kindled fire in the combustion chamber will heat the tubes; at the same time start the donkey pumping in water, which contracts the scale until it bursts and falls down wholesale.

172. Another way is to fill the boiler with steam from another boiler to expand the scale, which is assisted by opening up all dampers, fire-, and tube-doors. These expedients, like other desperate cures, are not to be recommended. But in all cases of blowing a boiler down in a hurry keep the donkey pumping into it all the time to prevent hardening of the scale.

173. WASHING OUT

of a boiler is very important, especially after any scaling. Every little bit of scale sticking between the tubes acts as an accumulator, either if it remains or drops down, say, on the furnace crowns, and may lead to serious consequences. The hose should be taken inside, and with good force from the donkey applied vigorously from underneath upwards to detach any ulcer barnacles about the quiet places; then from the top downwards.

174. After washing out, examine the quiet places for pitting, and wherever found, lose no time and scrape the parts well out, even to partial removal of the plate; drill them out a little bigger and deeper, then fill with Portland cement and Archangel tar mixed, rubbing this well into the pits, or, failing the tar, mix with sea water.

Blowing.

175. When opening a blow-off cock always keep the spanner or key in your hand until you have again shut it. Always leave a boiler clean and dry to prevent corrosion, pitting, and rust.

A. Use fresh water when you can get it good for the purpose. If it contains carbonate and sulphate of lime much in excess of sea water, it is unfit for boiler purposes. The salt in sea water will not harm your boiler if you do not tamper with it. Good sea water is always preferable to bad fresh water.

B. Scum as little as possible, as you are blowing the water off the surface which is fresh, although working with sea water (§ 35). You may prove this by testing from the gauge-glass and comparing salinometer results. Use discretion and choose the least of several evils.

C. Blow as little as possible, just sufficient to eject the solids; if you blow more than this, you are putting out what has lost its lime, and taking in what is charged with lime, which will deposit on the surfaces (§ 77). Much blowing results in scale. Water at $2\frac{1}{2}$ salinometer density without its lime is better than that at $\frac{1}{33}$ with its lime.

D. Be sure your feed is all returned to the boiler.

Never *freshen* from the scum cock.

Never test the density from a gauge cock.

Be sure your feed is not augmented by a leaky condenser.

Never run above $\frac{3}{33}$ density owing to *salt* deposit.

Look out for donkeys, winches, and glands robbing your boiler of its fresh water.

RETENTION OF WATER.

176. Coasting or river steamers are most likely to have scale if *hard run*; if not, they need have none at all. Say, a vessel leaving port filled with sea-water runs 12 hours; *banked fires*, 6 hours; waiting with steam up, 8 hours; *keeping steam handy*, 3 hours; running, 14 hours; banked fires, 10 hours; fires drawn and clean tubes and backs, 24 to 30 hours in port, steam winches and donkey going, no donkey boiler, only one main boiler. This is a job to take on scale.

177. When an ocean-going steamer requires to freshen the water for density there must be something wrong; there must be leakage, bad glands, or

loss of feed water. The auxiliary has been too much in requisition. If you take from the sea on an ocean voyage more than 3 inches of gauge glass per day there is something wrong. That fact should be obvious to the engineer long before it is necessary to blow the boiler. A boiler has been run 30 consecutive days without blowing, then the water was only $2\frac{1}{8}$ density. One could see no difficulty in running other 20 days as far as density was concerned, but another element had to be considered of nearly equal importance, namely,—

Acidity.

178. When working many days with the same water, acidity is quite a reasonable condition. If we bear in mind that the feed water with which we have been working so long is distilled or condensed water, having from the first lost that preserving property natural to water, viz. that of its fixed air. It has been pumped, generated into steam, condensed, pumped again, and so on, for many thousands of times undergoing this metamorphosis, so, after the manner of all liquids, it turns sour or acidulated. Milk will not stand much jolting; beer will not, but would soon be turned by it into vinegar.

179. Water deeply acidulated will attack naked iron fiercely, that is, malleable iron that has lost its original rolled or forged skin by having been chipped, planed, or turned, such as combustion-chamber stay edges, turned edges of stay nuts, and such parts

inside the boiler, also piston nuts, or rods near to the nuts, and necks of crank pins, the acidulated water escaping by the glands and running down. The action will lay bare the fibres of the iron, giving it in forged iron the appearance of filigree work. It will sometimes attack *several* boiler stays fiercely, while others will not be affected.

180. Experience proves that this condition of the water is not general in a boiler, but is concentrated chiefly near any spot sheltered from agitation. When in a concentrated form it will charge about the water surface as an "irreconcilable," inducing priming until ejected through the cylinders, when it will reform on return, and will probably at length be ejected by the waste-steam pipe. This acidity prevails in most boilers more or less, whether connected with surface condensers or not, as can be witnessed by the eaten-away state of valve spindles, and even cylinders of high-pressure engines.

181. It is also promoted by boilers not being thoroughly cleaned out; that little drop so often left in land boilers is the germ of this and other evils. It doubtless has the property, if in sufficient strength, of engaging the salt in a boiler.

182. At a certain stage of acidity the salinometer will (notwithstanding the additions of sea water) fall back, that is, the water will freshen. The presence of acidity can be proved before it affects the salinometer, by putting a small piece of iron (after having filed one face of it) in a pot with some of the feed water; the acid will show a finely engraved illustra-

tion of the *structure* of the iron as well as the process itself.

183. Changing water for acidulation must be done by the scum cock, the reverse of that for density.

184. The indiscriminate use of soda will induce acidity, therefore, soda should be used in moderation. Acidity is easily stopped either inside a boiler or out, by treating the parts affected with Portland cement as a thick paint, or changing water.

185. Soda,

as a preventive of scale and pitting, is a very safe and cheap remedy. Its efficacy in softening hard waters, preserving boilers and culinary vessels from rust, has been known from time immemorial, and its general applications are better known than any other substance.

The soda water used by iron-turners imparts a fine skin to the iron as it leaves the tool. This skin or film of soda thus put on with the tool is a preventive against rust, in all turned or machined articles of iron. This is well known and appreciated throughout the trade by master and man alike.

186. The properties of a solution of soda can be brought within every one's experience, as in the manufacture of common screw nails and other kindred articles, where soda is used with the tool or as a bath, whence they retain their pristine polish. This preserving effect of soda is secured by the articles being enveloped in a thin coating of its

substance, the cutting of the tool imparting the heat necessary to make it a fixture. In the boiler the same interposition of its substance takes place between the heating surfaces and the depositing solids, effectually baffling the process of depositing from the incohesiveness of the individual particles of solid matter as well as the protected heating surfaces; the lime salts can, therefore, only find rest in the bottom of the boiler. This property of the substance also serves as a preventive against rust, and deterioration of the plates inside, either from acids in lubricating oils or the water (§ 346), (which is another leaf of the book of domestic experience, §§ 133 B, 174), and prevents corrosion and general deterioration outside, " in so far as regards results from leakage"; the preserving quality being inherent in the soda-treated water when leaking out and wherever it goes. This property of soda is mechanical.

187. Another virtue we claim for it, as applied to the steam boiler, is a chemical property, viz. that of neutralising the acids in the feed water, and organic matters that find their way into a boiler, committing, in many cases, great havoc in the shape of pitting (§ 133). Soda being an alkali is therefore a natural cure.

The conjunction of the soda as an alkali with the acids in the water gives a double issue, inasmuch as the active energies of the acid and the latent force of the alkali, chemically combined and focussed, produce a natural issue as a salt; the acid is therefore gone.

In our case the alkali was a partial acid or carbonate of soda. Therefore, although producing no neutral salt, the following chemical exchange is thereby effected: sulphate of lime is decomposed by its means and precipitated as carbonate, while a soluble sulphate of soda is formed.

188. If we now look over the analysis of boiler scale (Fig. 7), we find that lime is nearly the only ingredient of scale found on the internal surfaces of boilers, and that in the form of *sulphate*. Lime in any other shape, such as carbonate, nitrate, &c., can be practically digested and disposed of by the boiler naturally, but in the form of sulphate, ordinary boilers have not circulating power to dispose of it, and it therefore abides the most obdurate and intractable of all kinds of scale. Sulphate of lime and boiler scale are synonymous terms. With no sulphate of lime in boilers, scale would never be heard of. In view of the property of this particular form of lime to adhere, the idea suggests itself—could the character of the salt not be simply changed? that is, could sulphate of lime not be changed to carbonate, or any other form of lime salts digestible by the boiler?

189. Now that is just what is brought about by the soda combining or *disputing*, as the case may be, with the sulphate of lime in the boiler. Thus—

Lime sulphate ⎯⎯⎯⎯⎯⎯⎯⎯ Carbonate.
Soda carbonate ⎯⎯⎯⎯⎯⎯⎯ Sulphate.

The carbon of the soda and the sulphur of the lime simply change places, giving carbonate of lime (§ 58), digestible by the boiler, and sulphate of soda (§ 47), perfectly soluble in the boiler. Our great enemy sulphate of lime is now disposed of. Under this same chemical process of exchange we have several processes, such as Clarke's and others, for treating feed water before it enters the boiler.

190. On again referring back to Berthier's table (§ 61), we find it proved that it is not only the character of the solids received firstly into the boiler, but the combinations and exchanges taking place inside, with which we have to deal—the physical and mainly the chemical changes and exchanges constantly at work during the process of concentration. We have column A, carbonate of lime, sulphate of lime, and no sulphate of soda. At column B all is changed, carbonate of lime and magnesia deposited or exchanged for sulphate, which has unaccountably increased by 700 per cent., sulphate of soda having now just come into existence. In the third column the lime sulphate is deposited (at any rate the lime part thereof), while the sulphuric acid part has found another connection, the sulphates of soda and magnesia being increased. The chlorides and sulphates in nowise maintain their integrity of volume throughout, even their combining quantities considered, owing to other trains of decomposition, apart from our business. We have, however, these corroborative proofs upon the face of the table itself, that carbonate of lime is deposited at an early stage, and that sulphate of lime and sulphate of soda are in reality manufactured inside the boiler while at

work; therefore an analysis of the feed water is no criterion for proportions of solids in the same water concentrated to $1 \cdot 140$ or $\frac{5}{33}$.

As sulphate of lime is created, so to speak, inside the boiler within the ordinary working temperature naturally from chemical contact with other solids, the inverse operation, that is, the dissipation of the salt as sulphate, induced by natural means, is as natural as it is for bane and antidote to grow together.

191. As Mr. F. J. Rowan read before the British Association at Glasgow:—"The action of soda ash or carbonate of soda, under these circumstances, is a very interesting one, though, perhaps, not well understood by those using the substance in this way. Sulphate of lime is decomposed by its means and precipitated as carbonate, while a soluble sulphate of soda is formed. The neutral carbonate of lime is likewise produced by reaction from the bicarbonate in solution, and, as thus formed, it will not adhere to the boiler surfaces, but separates as a loose powder or mud, which can be blown out of the boilers or otherwise removed as sludge. The use of too much soda is injurious in its effects, as the excess boils up and passes over in the steam to the cylinders and pumps, where it clogs the pistons, and otherwise interferes with the proper working, by making combinations with the oils and greasy matter employed in the machinery. The lavish use of oils and greasy matter intensifies the action where it is present, and it has been found that the carbonate of lime itself

has passed over from the boiler with the steam, and has entered into combination with the grease where enough was to be found."

192. The deposition of the carbonates of lime as an innocuous mass is promoted. Mr. Mallett, in a report on iron corrosion, says, "I may state here, upon the authority of my own researches of past years, that the carbonic acid is evolved, and the carbonate of lime begins to deposit, as soon as the feed water is heated to 190° F."; and further, according to an article by M. Coustie, translated by Mr. R. Mallet, if water containing sulphate of lime in solution naturally (whether fresh or salt), and which is to be used for evaporating into steam at any given pressure, be exposed previous to its introduction into the boiler to a temperature of say 271° F., which is that due to a pressure of $2\frac{1}{2}$ atmospheres nearly, it will at once deposit there almost the whole of the sulphate of lime; and if heated to 300° F., absolutely the whole of that salt.

M. Coustie's tables on boiler scale show that about 86 per cent. of it is sulphate of lime, which when once deposited is perfectly insoluble and as hard as natural alabaster, which in fact bears nearly the same chemical constitution.

The collective testimony of these and other authorities must be conclusive that lime in the form of carbonate, whether its deposition be accelerated by the soda or not, is a comparatively harmless deposit, and that it takes place during the raising of steam, and by the time steam is up it is all deposited.

However, it ought to be got rid of at the first opportunity (§§ 87, 164).

193. All are well aware of the efficacy of soda when used for washing purposes with hard, intractable water, or for the general purposes for which soda has been used from time immemorial, which manifestly includes its application to steam boilers, in which its efficacy against incrustation and the destructive pitting cannot be overrated.

The efficacy of the substance for good when used in moderation would also indicate an evil tendency if used to excess or improperly administered, which is certainly the case. If the soda is allowed to get beyond a certain strength in the boiler, it will mate itself with any lubricating oils (§§ 346), tallow, or fatty matter of any description that may have passed with the feed, forming a light curded scum on the surface, which will pass with the steam into the cylinders, where the rubbing and churning will at once convert the whole into an insoluble soap, which very soon clogs the pistons and packing springs into a solid mass. Of course in this way it neutralises all lubrication of the cylinders and valves, and produces the same pitting tendencies as described at (§§ 133, 134) as a result from common soap, and also incites the acidulation of the water. In short, soda used to excess promotes what its moderate use prevents.

But the greatest evil produced by this soap formation is, that it gets resolved into patches in the cylinders and on piston rods and valves, producing such a "rug tug" effect as to very soon cause serious break-down. This will extend to winches or all engines using this boiler. A boiler will always indicate its soda strength at any leakage or cocks, such as gauge-cocks and test-cocks where any lime crust gathers. When this crust gets a yellow tinge, no more soda should be used until the other is more diluted. This yellow tinge is a reliable practical safeguard.

194. Between the moderate and immoderate use of the substance as a remedy, there is a margin large enough for safety in all reason. But we must bear in mind, that although soda kills the acids already referred to, it does not prevent the acidity of the water, as described (§ 184).

195. The following may be taken as safe and proper practice for the use of soda. For every 100 gallons of sea water allow 1 lb. soda crystals, strength 48° to 52°. Soda ash is about half the strength of crystals, caustic soda is about three times the strength of crystals; therefore, with a bucket or other vessel used for the purpose marked in gallons, the salinometer used as a lactometer will tell the quantity of either of these kinds of soda for the desired strength as compared with crystals.

Before letting the water into the boiler empty this melted soda with a dish on the places most addicted to scale. When the water is put on, this melted soda will be borne upwards, anointing every corner and surface as it rises, most completely. To maintain the same soda strength in the boiler, add of melted soda each time the extra feed is used, in proportion to quantity of water, through the condenser, or by safety valve when steam is down. When soda is put in and no blowing down takes place, the substance does not deteriorate, but acts again and again.

Caustic soda cannot be well recommended, being too potent a substance in the event of mistake. It acts savagely on copper pipes and brass cocks. Good brass is necessary for all cocks about boilers where soda, especially caustic, is used. If the purity of caustic soda was as certain as its efficacy, it might be more recommended.

ZINC.

196. The use of zinc in steam boilers has been practised for a number of years, either as plates any convenient size, suspended between the boxes of tubes or otherwise as desirable. The benefit of having zinc in a boiler appears to have been matter of discovery as far as the plates are concerned. Many who use zinc for boilers simply do so because others do it, trusting blindly to the popular favour with which zinc plates are held. Others claim to have secrets respecting the use of it. It is, however, very curious that with most engineers who use zinc, opinions respecting its practical merits are very much divided.

197. From my own experience I believe in its efficacy as a mechanical interposing substance, similar to that of soda; but if having to choose

between the two I would prefer the soda. I would also allow for zinc the property of galvanic action pervading the internal surfaces, similarly as soda, to baffle the action of deleterious acids and pitting. This property may well be taken for granted, as the water in the boiler itself, being a partial acid, becomes the solution, and the boiler itself the battery as well as the subject of operation,—the copper pipes and other copper gear inside acting as co-agent with the zinc in the production of the galvanic action; and as the acidity of the water increases so does the efficacy of the zinc plates. But without the zinc being protected by an amalgam it is not feasible that such a protracted and extensive action could be sustained. There is less doubt of its chemical virtue (so far as it goes) as an alkali making exchanges producing salts, soluble for insoluble, as zinc sulphate.

198. But whether the zinc alkali intercepts the sulphuric acid in combination with the lime, as sulphate in suspension, allowing the lime thus to be taken up by the excess of carbonic acid and deposited as carbonate; or whether the sulphuric acid of the zinc sulphate in suspension be taken up by the zinc forming sulphate of zinc, which is soluble, also acting as an interposing agent over the naked parts of the boiler, if the lime sulphate be taken up by the excess of carbonic acid, and deposited as carbonate of lime; or whether the salt as formed be carbonate or sulphate of zinc, or be otherwise combined—is not plain. The whole subject of the use of zinc requires more light shed upon it altogether.

This much, however, has been proved in its favour by the use and disuse of it, that it is effective in protecting the plates from the action of acids, either from the feed or cylinder lubricants, and pitting, and that, conjointly with soda, it acts beneficially in preventing scale.

199. At any rate, their action is proved by their rapid decomposition, as also the decomposition of any ironwork, such as a pipe hanger bearing a copper pipe. If a piece of sheet-lead or other non-conducting substance be not put between, the iron, if malleable, will be eaten away in a surprising manner.

These are the only bad effects resulting from the use of zinc, which, being known, are easily abated. The theory is open that a potent chemical base such as zinc, in presence of heat and acids, is bound to be a power for good or evil, and effect there must be. If no ill effects are experienced, then, to say the least of it, the use of zinc is safe practice, and may avoid reflections.

200. Another form of zinc in use for boilers is balls from 18 to 56 lb. weight, with a copper core from which a wire is led to any part of the boiler, or combustion chamber-top or furnace, and soldered. This is a patented arrangement. The patentees assert that when scale exceeds a certain thickness on a plate, it is curled up by the galvanic or chemical action, and peels off. But with the use of zinc in any form comes the question of expense. In a case of opening up two boilers with two balls (or electro-gams, as they are called by the patentees) in each,

in one boiler both balls were decomposed and gone, while in the other both were intact. Both boilers were working under the same conditions, as were also these zinc balls. There was no appreciable difference in the effects produced between the two boilers.

201. We know that in ordinary galvanic batteries of zinc and copper plates the zinc is treated with a metallic amalgam, otherwise it would be decomposed very quickly with even a very weak solution. On the other hand, the plates treated with this amalgam suffer no deterioration more than the copper, while the galvanic action is not sensibly diminished. Therefore, so far as the galvanic virtue is concerned, the expense of zinc could be made one only of first cost by so treating the plates or balls. At any rate, the expense could be reduced to a minimum. But the mechanical virtue claimed for it would in this way be destroyed, as the diffusion of its substance would be stopped.

202. The zinc in boilers is supposed to be oxidised, and the hydrogen, the other element of the water decomposed, is assumed to accumulate on the face of the iron and keep the deposit of scale from adhering. By some it is believed that a galvanic tremor pervades the face of the plate. It would appear that any physical effort extracted from the zinc could not sensibly affect the amount of water evaporated for any appreciable time.

In using zinc plates, 12 inches by 6 inches by 1 inch is a very handy size. Several should be put in sheet-iron boxes or pockets, perforated with a number of holes and suspended from the stays down to where wanted, say near the furnace crowns or any part most in want of them. They must be secure from knocking about, and must be kept clear of everything. Heavy plates are sometimes used with a cast hole, by which they are suspended. This is not a good plan, as when the zinc gives way when decomposing it is practically lost by falling down amongst the mud in the bottom. The boxes or pockets referred to greatly favour economy.

203. Another very pernicious practice is to fix such pockets to the sides of furnaces. This is very dangerous, as, thus situated, they accumulate any falling scale, and by thus preventing the access of water to the plate, may bring about serious consequences (§ 208).

204. It has been found good practice on some railways where hard and soft water were available, to ring the changes on bane and antidote by putting the locomotives on either water alternately, which amounts to another very beneficial chemical exchange. And were all other things equal, this would be the sovereign remedy for the numerous ills that steam boilers are heir to. The next best mode of securing water for boiler purposes is to impound all rain-water available for mixing with other water, hard or soft, but in particular to fill up the boiler from time to time.

Scale Prevention.

205. In the use of compounds for the purpose of counteracting the effects of substances in the water great care must be exercised, inasmuch as we have double issues to look out for. Anything we put into a boiler to do one thing will, in spite of us, do two or more things: the first may be what is wanted correctly enough; the second or reactionary effect may be very mischievous, because if being an acid and engaging alkalis in the boiler, it may bring about several courses of decomposition until neutralised; then the question is, What of the resulting salt? Is it good or bad for the boiler? The same question, but inverted, arises when the compound used is an alkali, as there is no saying what may be brought about with altered conditions of even well enough known substances inside of a boiler. But any substance whatever that would by any train of decomposition or interposition clarify the water of suspended matters, notably lime sulphate, by promoting either its solubility or deposition by direct chemical exchange or combination, without acting injuriously on the boiler, at a nominal outlay of trouble and expense, is a desideratum long wished for, be it nostrum or not. It is unnecessary to refer to the numberless alleged cures for scale, as up to the present time no substance has so faithfully fulfilled the requirements as soda for efficacy and economy, and it has been employed for the purpose from times remote.

206. Salts, acids, or gases will not disappear in a

boiler without being blown out either with the water or steam. For scale softening, see § 166; scale removal, § 271; incrustation, §§ 36-42; scale chipping, § 272.

Boiler Medicines.

207. We may safely put it as proved that any substance whatsoever put into a boiler, in the first place does harm, because it thickens the water, thereby raising its boiling point, and also adheres to the heating surfaces.

Then the question is, Do they do any good besides the harm? If so, and if the balance be in favour of the good, the result will be in favour of the substance. Hemlock, oak wood, sawdust, tanbark, willowtops, logwood, leaves of trees, &c., all rate according to the amount of tannic acid they contain. They are all alike astringent and act on the iron plates of the boiler, certainly preventing scale more or less from the action of the acids. But the drawback is to get rid of the spent substance, which afterwards floats about the edges in knots, until taken up by other free acids, when it adheres and attacks the iron in shape of pitting, unless being scummed off with the water, blown off with the steam, or otherwise neutralised.

Now we can neither depend upon the blowing off nor the scumming being effectual, and we cannot afford the loss of so much fresh water, and that it be neutralised means that a new connection springs up, phœnix like, which in turn produces other issues or combinations to which there is no end. The first issue might be beneficial, the second harmless, as also the third, but the fourth may prove a greater evil than the original.

The administration of medicines to the human subject may well guide us. No doctor will prescribe medicine so potent as to attack and destroy the coating of the stomach. He will also provide that whatever substance is put into the stomach, shall in due time be got rid of. In the boiler case, this being more difficult and more a question of expense, is met by making equable exchanges (§§ 188, 189).

208. Serious injury to boilers, as also explosions, have often resulted from the indiscriminate use of anti-scaling compounds; as if frequent blowing is not carried on these spent substances, lying inert on the heating surfaces, will act similarly to scale, and bring about the same consequences.

But when the compound used deposits on the surfaces as a soapy or glutinous matter, the consequences then involve more than the collapse of furnace crowns or buckling and burning of the plates (§ 276).

The excessive use of cylinder lubricants, as oil and tallow, is another serious aggravation, and if we add to this, waste and old bags, that are taken into boilers by scaling boys and left there, the conditions necessary for a serious explosion or collapse of furnace, as at § 309, are complete. It is not once or twice only that serious consequences have resulted from an old bag being left in a boiler.

Water Expansion.

209. We may have a better conception of the deposition of the solids in water by viewing it as a mechanical phenomenon. For instance, a tumbler of mineral water, however pure it may appear to our eye, will be proved to our own satisfaction by our senses of smell and taste to be strongly impregnated, notwithstanding its purity, which proves that there is a *bonâ-fide* solution. On the application of heat, these solids in the water will be suspended in turn, the water will have lost its clearness; this is simply the solids finding their way to the bottom or depositing. This falling solid matter must therefore be heavier than the water *is now*, bulk for bulk. Therefore the term "specific gravity" will simply represent the weight of the solids added to that of pure water (§ 97). The following table shows how water in a boiler expands while raising steam. Two feet in depth will expand $\frac{7}{8}$ between 32° and 212°, equal to $\frac{1}{3858}$ of its own bulk for every degree of heat. Boils at 212° at sea-level, maximum or greatest density 39·38° Fah., or at the rate of about $\frac{7}{16}$ per foot. Therefore a boiler with 12 inches range of glass, and at $\frac{3}{4}$-glass with cold water at 32° will be out of the glass before steam is up *by swelling with the heat*.

$2\frac{1}{2}$ feet	expands	$1\frac{1}{4}$	inch.	7	feet expands	$3\frac{1}{16}$	inches.
3	,,	$1\frac{1}{2}$	,,	8	,,	$3\frac{1}{2}$	,,
$3\frac{1}{2}$	,,	$1\frac{5}{8}$	,,	$9\frac{1}{2}$	,,	$3\frac{7}{8}$	,,
4	,,	$1\frac{3}{4}$	,,	10	,,	$4\frac{3}{8}$	,,
$4\frac{1}{2}$	,,	$1\frac{7}{8}$	,,	11	,,	$4\frac{3}{4}$	,,
5	,,	$2\frac{3}{16}$	inches.	12	,,	$5\frac{1}{4}$	,,
6	,,	$2\frac{5}{8}$	,,	13	,,	$5\frac{5}{8}$	,,

Water Expansion.

210. It requires more heat to raise 1 lb. of water one degree than it does to raise almost any other substance. The heat required to raise 1 lb. of water one degree will raise 9 lb. of iron and 30 lb. of mercury.

Water is formed of oxygen, 16 parts; hydrogen, 2 parts; weight of a cubic foot, $62\frac{1}{2}$ lb.; of salt water 64 lb.; weight of column 1 inch square and 1 foot high, ·434 lb.; 1 cubic inch of fresh water evaporates into 1700 lb. of steam. Water can bear steel wires carefully laid on its surface if not agitated so as to break the skin.

Water can be cooled several degrees below zero if not agitated.

Water is incompressible.

In rain-water chalk is partially soluble.

Rain-water and melted snow do not taste alike.

Water expands $\frac{1}{24}$th between 60° and 212°.

Water is not sensibly expanded by solids in solution up to the point of saturation, after which the water increases in proportion to the bulk added; whether held in suspension or deposited, the relation is only mechanical, while the former is chemical.

In practice with sea-water the salt left behind by evaporation while in solution only fills up the minute interstices between the particles of water until we reach $\frac{4}{3}$rds.

211. On referring to the expansion table we find water is at its greatest density at 4° Cent., increasing in bulk as it is heated until it gives off steam, and

also increasing as its solidifies into ice, at the rate of $\frac{1}{17}$ of its own bulk. This accounts for the bursting of pipes, pumps, &c., containing water during frost. For instance, a propeller boss being full of water on a ship going into dry dock will burst under severe frost.

EXPANSION OR DILATATION OF WATER AT ORDINARY TEMPERATURES.

	Centigrade.	Fahrenheit.	
Freezing	0°	32°	1·00012
	2	35·6	1·00003
	3	37·4	1·00001
Greatest density	4	39·25	1·00000
	5	41	1·00001
	6	42·8	1·00003
	8	46·5	1·00011
	10	50	1·00025
	15	59·3	1·00082
	20	68·2	1·00169
	30	86·4	1·00420
	40	104·4	1·00766
	50	123	1·01189
	60	140	1·01672
	70	159	1·02237
	80	176·8	1·02811
	90	195	1·03553
Boiling	100	212	1·04312

Similarly, if a vessel be filled up with water, and enclosed when heat is applied, it will expand and burst the vessel notwithstanding its strength. Now this increase of bulk by heat must make the hot-water lighter, bulk for bulk, than the cold; therefore this expansion means also decrease of specific gravity. The increase of bulk can only be a *bonâ-fide* swelling of the pure water itself; the solids held in solution while the water was cold cannot sensibly contribute to this phenomenon, disparity of volume and expansive properties both considered; and if water when cold is 1 in 24 less in bulk than when hot, this increase of volume must mean also decrease of substance, decrease of strength, or inability now to bear the solids in solution as before; and assuming the affinity of the water for the solids and the solids for the water to be at *par* when cold, it follows as a matter of course that the solids will be deposited in turn with the increase of heat.

EXPANSION OF BOILERS.

212. The expansion of iron when heated is also a very important property affecting steam boilers, inasmuch as the unequal expansion and contraction of the different parts of a boiler at the same time, such as the steam space being heated to 300° while the water space is at 214°, means wrenching of the joints and general fatigue, and TEAR AND WEAR. This constitutes the predisposition of high-pressure boilers to leak, and must in many instances be a contributory cause of explosion. In a very long Cornish boiler the unequal expansion between the top and bottom part amounts to nearly two inches—that is, with the ends resting the centre will be arched to about half the thickness of brick on flat (§ 301). This unequal expansion is greatest with high steam and drawn fire.

213. Latent Heat

of bodies is, to the uninitiated, a most remarkable property, an apparent paradox. This property of water and steam is indeed only known by name to the bulk of those who ought to know better, and the neglect of the subject, or ignorance of it, has resulted frequently in serious consequences.

The sketch (Fig. 13) is intended to make a simple illustration of Joule's equivalent of heat, viz. that in the latitude of Manchester 772 lb. falling a height of one foot, or 1 lb. falling a height of 772 feet, would in effect be equal to raising the temperature from 50° to 51° F., or one degree. Now let us suppose the large and small balls in the sketch to represent a large and a small boy come to see-saw on the plank which they find in the position A. They with apparently a common instinct adjust the plank as at B, and get on to it; as quick as thought they readjust the plank as at C, when they succeed in having see-saw. These boys adjusted the plank in the hundredth part of the time it would take a philosopher to demonstrate the theory. C a represents the small boy with the long leverage, as the one lb. weight with the long drop; b represents the big boy with the short leverage as the 772 lb. weight and the short drop. The effect of the blow delivered to the stob c by the small boy will be the same as that delivered to d by the big boy; both stobs will be driven the same distance into the ground, each blow would represent a *thermal* unit or degree of work corresponding with the *thermo*meter. Let us now follow this up into the latent heat of water and steam (Fig. 14, §§ 221, 222).

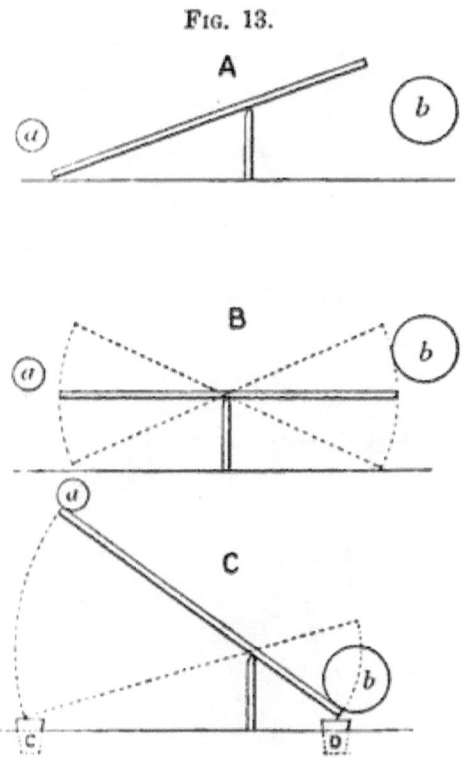

Fig. 13.

214. The lines headed P, T, L, represent respectively pressure, temperature, and latent heat. Down the line of temperatures at the bottom we come to absolute zero, 461° below common zero. Now this unearthly descent into the nether regions is a supposed point, where a body has absolutely no heat, and, indeed, where there is absolutely no body

to receive heat, but is held as a computative limit of useful effect. Here again we find the small boy, *a*, and big boy, *b*, of the sketches; the small one falls

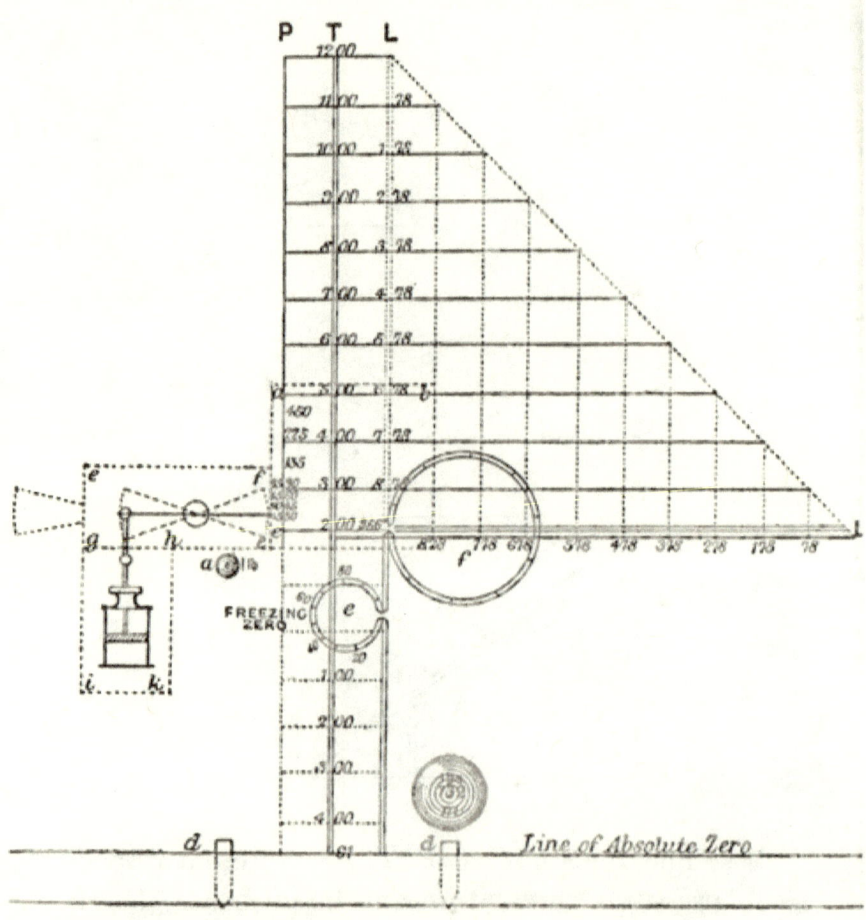

772 feet, driving the stob in one inch, the large one only falls one foot, and drives the stob *d* into the ground one inch also. This then is Joule's equivalent

of heat, or degree of useful effect. Suppose now that the line L was a glass tube, as shown with a loop, and afterwards going off at a right angle to L s, and that T was also a tube, both filled with mercury and heated from below equally. Now let T and L ascend their respective tubes together. They will keep abreast of each other all the way, but on reaching 32° above zero L suddenly leaves the track and disappears into the bush; T instantly stops, both having travelled steadily 493°. After making a circuit of 143° L appears just where he lately disappeared. They again walk in company for 180° when L again disappears; T instantly stops as before, while L makes another circuit, this time of 966°, reappearing where he disappeared just as before. T, who has been waiting all this time, at once resumes his journey in company with P. L being *worn out* and unable to make further headway, remains for their return. P soon leaves T behind. On the return journey P and T arrive together at the place where they had left L, who, on their arrival, retraces his steps round the loop, T again waiting for him, while P retires to his own place (§§ 215–16). The first disappearance of L was at 32°. On looking at the diagram we find he had been walking round the loop steadily until he returned.

215. Put some ice into a bucket and put it on a fire. As soon as it melts sufficiently put in a thermometer; it will stand at 32°, and will not move until the last bit of ice is melted. It will then expand. Because the water was kept cold by the ice melting,

and as ice cannot exist above 32°, it must be plain that the thermometer could not rise while any ice existed. Therefore T had to wait at 32° until L returned, or, the thermometer stopped at 32° until 143° of heat had passed into the bucket to melt the ice. The heating was going on all the time as the figures on the loop will show. This 143°, or the latent heat of water, is insensible to the thermometer, just as we have already said. The thermometer must remain stopped while any ice existed. This interregnum is not so important for our purpose as the next disappearance of L at 212°, or the stoppage of the thermometer again. A. Water at 212° changes its constitution, resolving itself now into globules, the inside of these globules being cold. Again, let us fill a pot with green peas and put it on to boil; when it starts boiling the water will be at 212°, so indeed will the outside husk of the peas, which will now be beginning to burst, shedding their cold contents into the water; and the solid contents of a pea when turned out is many times greater bulk for bulk than the husk, and therefore takes off the heat.

B. Again, who has not found a potato on his dinner plate quite well done outside, and the inside yet hard and raw? Now the outside of the potato was cooked at 212° of course, and the inside would have been *by and bye*, had the pot been allowed to remain on the fire.

C. Again, in the ice case we had the fire heating the water while the ice was busy cooling it. In this case the fire is heating the water too, while

Latent Heat.

the cold contents of the potatoes or peas are cooling the water as fast as it is heating. The very same process takes place with globules or atoms of water.

216. By the diagram (Fig. 14), we find that on the thermometer T reaching 212°, it remains, while the heat is directed to the internal heating of the water. This done, the mission of L is accomplished, and he remains at 212°. Meanwhile P pressure rises and takes up the walking (pressure rises in a greater ratio than heat). Again, as steam falls as on the return journey to 212°, P, or pressure, ceases to exist. The thermometer T again remains stopped, while L retraces his steps around the loop, or, in other words, while the heat previously stored up is dissipated.

From 32° to 212° or 180° cost 180 lb. of coal; 966° will therefore cost 966 lb. of coal, or $5\frac{1}{2}$ times that amount. Having now 15 lb. of steam by the steam gauge or pressure line, which is now our guide, the water, if clean, does not rise above that temperature, as the steam does (Fig. 14, T P). The question now is—What of this 966° of heat stored up latent in the steam and the economy of the boiler compared?

217. Five and a half times the amount used for heating the water expended on latent heat is just 550 per cent. sunk on the latent heat. Although it can be recovered theoretically, as the steam is again turned into water, for steam purposes it is lost. And this heavy expenditure each time steam is

K

raised in a boiler tells very heavily against such as are cooled down and lit up often.

With boilers working continuously, such as marine boilers, for say 550 hours at a spell, this makes the average outlay on latent heat as 1 per cent., whereas an engine engaged say two hours per day, has an average of 2·75 per cent. on latent heat as loss.

218. The best reckoning for this expenditure is to allow as much coal as required for each lighting up, independent of the work to be executed by the engine.

This outlay need not be confounded with what benefits there may be in working at high pressures with the latent heat increasing, so to speak, while the steam is falling. The latent outlay stands by itself, like an auctioneer's licence, and must be met before a blow can be struck or a hammer raised.

The smith has to meet a similar outlay by having to heat his iron, and according to the amount of hammering he can bring to bear upon the hot iron will his success be. It is a good old precept for all to strike the iron while it is hot, therefore, when steam is raised, the most should be made of it, and the lighting up should be reduced to a minimum. Fires should be banked at night for next morning, provided that the water for supply is not productive of scale, but otherwise it should not be done. Circumstances in such cases require discretion. If high-pressure steam could be got easily, there would be no doubt of the advantage, but when steam has for its base boiling water of 212° only, the natural consequence is, this face of water controls the temperature of the steam by conduction (§ 227).

Recapitulation.

219. On the line of absolute zero we find the stobs h, n, and the balls g, m, each falling their respective height, illustrative of Joule's equivalent of heat, viz.:—A pound weight, falling 772 feet, is the same in effect as 772 lb. falling 1 foot; each strikes the same amount of blow, viz. driving the stobs each into the ground 1 inch, which Joule puts as a measure of heat, or one degree of heat.

Upon this rule all calculations pertaining to results of heat and work are based. Rising then from absolute zero, each step represents one degree or unit of work.

At 461 the stoppage of L to deliver 143° of heat is necessary to melt the ice (§ 215); this heat is not shown by the thermometer, and is termed insensible, invisible, or latent heat, and stands for reckoning units of work. This done, the thermometer rises to 212°, when it again stops, this time to deliver 966° of heat, this amount being required to turn boiling water at 212° into steam at 212°, or in other words, to *evaporate* the water (§ 215, A.) At this point the latent heat stops, thermometer rises, and pressure begins; and as the thermometer or sensible heat increases, the latent decreases (Fig. 18). This 143°

shown as the first loop as latent for water, and this 966° latent for steam, describes the whole question of latent heat. The first loop e affects us during times of frost to a trifling extent only. The second, or steam loop f, is very important, being like a reservoir we fill at starting to the top, and lose three-fourths of it every time we stop. From L to L s, is 966°, or simply the loop straightened and shown vertically. L l to L s is the same shown horizontally. The triangular figure L, L l, L s, represents the constantly varying relations of P, T, L in practice ; 1200° of heat from common zero as a maximum limit for line T is credited to T, 1178 being the point at which the latent heat is all impounded (see diagram), as shown at L and L s.

But comparing the ratio at which heat pressure rises (see P, T), we can see that the highest steam pressures ever attempted are far short of even 400° temperature, and a corresponding pressure for 1200° impossible. However, this point 1178 is reached previously by latent heat computatively, beyond which we may put it down as an arch-replicate of nature beyond our knowing.

As pressure rises, and consequently the temperature, the latent heat falls (Fig. 15), pound for pound according to Watt, but varying according to Rignault (§ 225.)

221. Fig. 15 shows the relations of pressure, temperature, and latent heat, or P, T, L enlarged from space, marked a, b, c, d on diagram (Fig. 14). Thus the steam at 30 lb. pressure on P, is 250° on T,

and 928 L. 135 P = 350 T, and 828 L. The line O P shows the relation of L to T (or, as it is termed also, sensible heat), as an angle of 45° of the one to a vertical line of the other. The loop to the right, also enlarged, shows the theoretical journey of L.

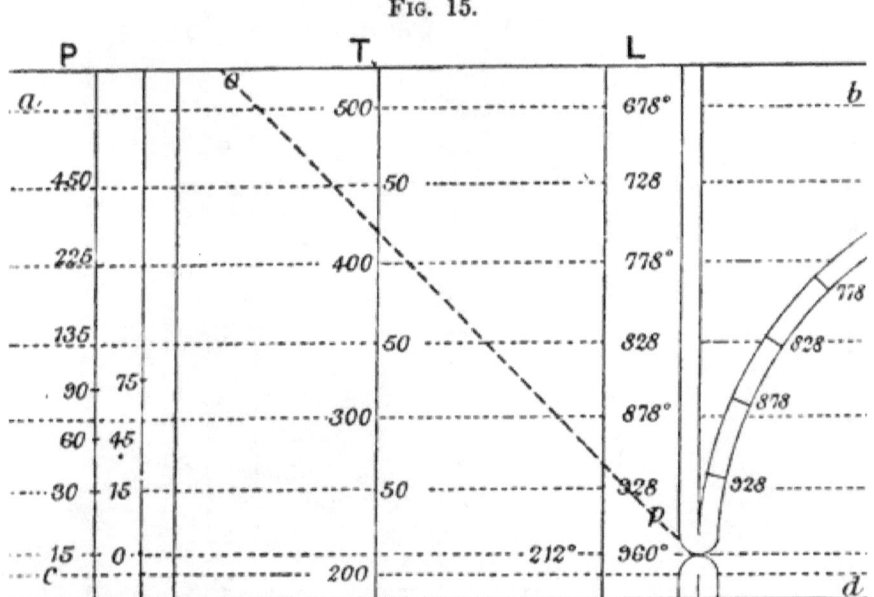

Fig. 15.

On the side of pressure the first column is pressure above the atmosphere, termed gross pressure (§ 348); the next column, from O upwards, is the pressure shown by steam gauge, and always 15 less than the other column; 15 up to 450 is gross pressure, O to 75 gauge pressure.

222. Fig. 16 shows how 1 cubic inch of water, a, when turned into steam expands into 1669 cubic inches, filling the entire space a, b, c, d.

134   *Steam Boilers.*

At 30 lb. pressure it expands about $2\frac{3}{4}$ times, filling the space a, e, f, g. In practice it may be

FIG. 16.

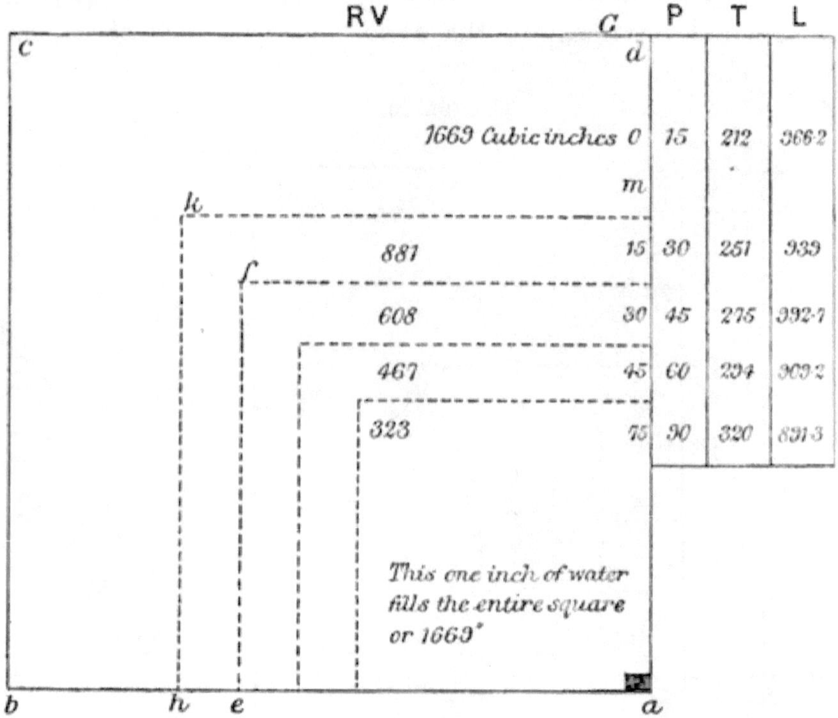

R V	G	P	T	L
1669 Cubic inches	0	15	212	966·2
881	15	30	251	939
608	30	45	275	992·7
467	45	60	294	909·2
323	75	90	320	891·3

This one inch of water fills the entire square or 1669°

taken that 100 cubic inches at 15 lb. equal 50 at 30, or 25 at 60. In all varieties the heat maintains its ratio.

One pound of steam at 212° =   180° sensible heat.
   ,,   ,,   condensed at 32° gives out   966° latent   ,,
                                          —————
                                          1146° total heat.

   ,,   ,,   at 240° =   208° sensible heat.
   ,,   ,,   condensed at 32° gives out 938·6° latent   ,,
                                          —————
                                          1146·6° total heat.

Recapitulation. 135

223 Fig. 17 is a practical application of the same subject. The terms "atmospheric pressure" and "gross pressure," are often very misleading, because we can only have this 15 lb. of atmospheric pressure after having created a vacuum to this amount. Therefore, in all dealings with high-pressure engines, gauge or actual pressure is all we have to deal with. In practice, the gauge pressure is always quoted (§ 21, A), and the vacuum, if any, added to it just as if it were pressure. This simple method saves much trouble.

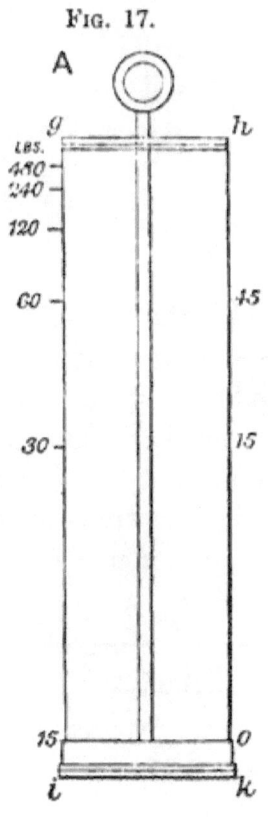

Fig. 17.

Again, this 1 inch of water referred to at § 222, A, evaporated into steam, equals 1669 inches at O (or steam gauge) pressure filling the annexed cylinder. Taking the handle h, and drawing the piston half-way to $30 = 15$ gross, the pressure will be increased to 15 lb., and by again halving the distance, stopping at 60, it will be 30 lb. (or doubled), and so on doubling each time we halve the distance, the temperature all this time increasing with the compression.

224. Referring this last illustration back to Fig. 14,

e, *f*, *g*, *i*, we find that as the piston in the cylinder D is moved with the pressure of steam, so does the record of sensible heat *f*, *c*, increase simultaneously

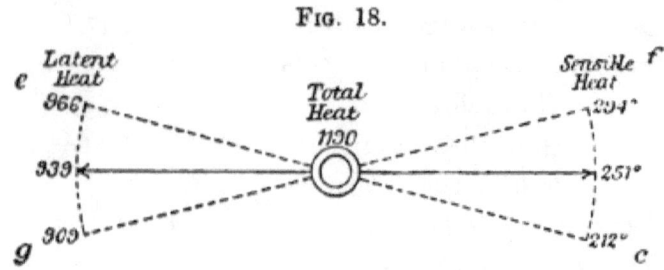

Fig. 18.

with the decrease of latent heat; e. g. while the total heat—that is, the figures at each end of pointer added together—is always a constant number (§ 225) (Fig. 18). On further referring to Fig. 14, we find that at a temperature of 1178, all the latent heat expended of 966° would be recovered but for the other matter of superheating (§ 227.)

225.

Gross Pressure.	Sensible Heat.	Latent Heat.	Total Heat.	Relative Heat.
	°	°	°	
15 lbs.	212	966·2	1178·2	1669
30 ,,	251	939	1190·0	881
45 ,,	275	922·7	1197·7	608
60 ,,	294	909·2	1203·2	467
75 ,,	309	898·5	1207·5	381
90 ,,	320	891·3	1211·3	323

It may safely be said that of all those engaged on shore, directly or indirectly in the raising of steam, not more than one in ten knows, or cares to know, about this remarkable and very important property, although all have realised its existence, many putting down the unaccountable *hiatus* to bad coal, &c.

Here is an illustration. A man has to raise steam in a boiler; after some time he feels the water hot, calculates getting steam in 15 minutes. On his return, notwithstanding a big fire, finds water same heat as before—no steam; half blaming himself and somebody else, he heaps on a still larger fire, allowing 20 minutes for steam to show, but in 5 minutes he is brought to the spot by a loud roar of steam from the safety valve. The man, although at his wits' end to account for this, has not sufficient confidence in his own senses to convict himself, or to make investigation for his own future benefit, or sufficient interest in the subject to make his experience known. He had been deceived by the property of the latent heat of steam which was unknown to him. How often have attendants been trotting from engine-room to stoke-hole back and forward, wondering why there is no steam, the water being so long hot! And then, after steam does show its rapid rise in the steam gauge, this is put down as a property inherent in the boiler itself, that she soon goes up after she gets a start.

226. To those accustomed to individual boilers, so much time is allowed from lighting fires to getting of steam, and with a safe margin allowed there is less source of danger. On the other hand, we have a great crowd of boilers of all descriptions, many

indeed of a class nondescript, with individuals attending them equally nondescript, having no qualifications for the job either from experience or aptitude, young boys in very many cases doing these duties. The boilers are not timed. Instead of having, say, three hours to get steam, many a time, owing to the attendant having overslept himself, the steam will be jerked on the boiler in half that time. Sticks, mats, bags, old furniture, &c., are thrown into the furnace; this practice involves other questions besides that of latent heat. By this vicious practice no doubt many of these terrible and unaccountable explosions occur, that happen just after steam is got up or just after the engine has been started. It is quite reasonable that a boiler can have super-generated steam from the concentrated flame acting so fiercely high up in the boiler where there may be only a small interposition of water, with the effect of a blow-pipe, thus so to speak driving the water off the heating surface generating steam perforce from the concentrated flame on the upper part before the under part has received its latent heat, or indeed before it is warm at all (§§ 296, 333). And if so much life and property is sacrificed or placed in jeopardy, owing to the subject being ignored, surely it is time it was brought forward.

227. Superheating

is best carried out in a separate chamber about four times the size of cylinders.

Using steam at a high pressure means also using steam at a high corresponding temperature, from 300° to 400° (table, § 225). This cannot be maintained in the presence of water at 213°, therefore this vessel is accordingly placed about the foot of the funnel, inside or by independent flues. Superheated steam is sore on packing, cocks, &c. The practice of superheating is doubtful economy, the advantages and disadvantages running nearly parallel.

Boiler Filling.

228. In filling up a boiler with water, one-half to three-quarter glass is safe practice. Boilers, as a rule, steam best with the water low, which, if practised, means great care. Methodical and regular feeding is a necessity. Running the water up out of the glass at the top is just as reprehensible as running it out of the glass at the bottom. In marine boilers with two or more working off the same feed pump, the water level of the gauge-glass is an all-important matter. In filling up a boiler the water ought not to exceed three-quarters of the gauge-glass, as owing to the expansion of water by heat (§ 164), when steam is up the water-level may be out of the glass at the top, then the appearance of the gauge-glass is similar to that found when insufficiently filled. This is very apt to deceive the attendant, and lead possibly to disaster (§ 267).

229. The Gauge-glass

is a very important part of the boiler mounting, and may be depended on generally to give the true level of the water, provided the water and steam passages be alike free. To ensure this the under cock should be opened at intervals of from one to two hours, and the water blown out of the glass. At less frequent intervals the water cock should be shut while this one is open. This manipulation will show if the steam passage is clear. Then opening the water cock and shutting the steam or uppermost cock will show if the water passage is clear, the under or stop cock being during this time open. If any stoppage occurs in either of these manipulations, we will have a case of the boiler showing

230. *False water*, which is a very alarming matter if the attendant does not comprehend the cause, and points at once to something being wrong. Indeed, the boiler may be in imminent danger, and to ensure against such as this, which might only be a false alarm, we must keep the cocks and passages clear and clean by using them frequently, as with the water passage choked, and the water in the glass discharged, no water will be shown at all; if partially choked, the glass will show above the true. Consequently we could have the water in the gauge showing, say, half glass, when in reality it was below the furnace crowns if a land boiler, and below the combustion chamber-top if a multitubular, involving very serious consequences and danger. Another fatal

trap showing also false water is a *mud line* on the gauge-glass. Suppose a boiler's water is at half glass in the gauge, and that it starts priming, which means all the muck at the bottom being mixed up with the water. It loses, say, 6 inches of water, or is just out of the glass at the bottom. Now we will have a mark in the gauge-glass at the level where the boiler started priming. This mark or mud line is apt to deceive the attendant, so that he takes it for the water-line. Such a deception may also involve serious consequences. Another matter of anxiety and annoyance is when the gauge-glass resolves itself into a

231. *Mud trap*, within the region of steam space there being always a light crisp scum worked up off the water surface with the steam. Nearer the centre of the boiler this scum, for the most part being the residue of cylinder lubricants or other substances, naturally descends in the eddy of the shell, and betakes itself to any orifice or recess in the steam space, similarly as heavier substances betake themselves to any orifice or leak in the water space, whereby to escape. In many boilers this scum becomes a serious annoyance, as when the gauge cocks are being used so as to blow the water or steam out of the glass, a circulation is thus produced through the gauge. This scum takes at once to the steam orifice and enters the glass, which then becomes a veritable mud trap. Often hours of manipulation of these cocks will not give you a clear glass. This nuisance is owing to the upper or steam passage being within the flush of the shell inside.

An effectual remedy is to insert an internal pipe from the orifice leading to the crown or steam space.

By the use of the gauge-glass we can also determine the amount of water we are evaporating in high-pressure boilers, and the amount of feed water we lose each day or watch with surface-condensing engines.

232. BOILERS LIFTING THEIR WATER

is due to the water being propelled upwards by the circulation from great local heat, chiefly furnace crowns, overcoming the weight of water, often termed "*boiling the water off the plate*," which is also a phase of priming. A contributory cause is boilers being overtubed. Boilers having this tendency take on scale on furnaces. As this is a kind of priming the same cure applies (§ 109).

233. FURNACES

determine the style of firing. They should never exceed 5 feet 3 inches or 5 feet 6 inches at most in length, so as to be manageable by an average man. When very long the back can neither be effectually reached with tools or coals, or even seen. The consequence is that the bars are allowed to run bare; air in excess is admitted, which is as fatal to proper combustion as insufficiency (§ 30). Besides, even with the bars covered, the air to feed it underneath is of little value so far back, being spent. Indeed, no greater smoke nuisance need be than abnormally long furnaces, and a long furnace also means a narrow furnace, which is just like a fire kindled in a tube. You cannot get sufficient air admitted at one end, and have to put up with nothing but smoke at the other, and smoke passing not ignited is lost.

A furnace is best whose length and breadth are nearly the same as in most locomotive and iron furnaces, and which has the bridge to suit the fuel regarding height.

FIRING.

234. Owing to the situation and surroundings of different boilers, as well as construction, there is generally a bit of trickery regarding firing, but the quality and temper of the fuel are the first consideration.

We have then to consider the manipulation of the fires, and this is best demonstrated by a good and an indifferent fireman relieving each other. The one will get his steam easily, the other will not, owing to his less experience or intelligence respecting the duties.

235. In *coaling bars* the furnace should be clean and clear of all clinker about the bridge. The bars should be covered lightly all over, but heaped up at the front, and a fire kindled in the ordinary way in front and under the heap. When this heap is ignited cover the bright parts with fresh round coal. Green coal should be put on in front, and pushed back every second or third firing, which keeps the fire bright at the back, thus consuming the smoke from the green coal in its passage across. The fire ought not to

clinker on the bars, as that impedes the passage of the air necessary for combustion, but should be pricked or eased up off the bars to break the caked fuel.

To keep *steady steam*, fires must be regularly cleaned at intervals of from 8 to 24 hours, depending on the quality of coal and amount of steam required in a given time with the facilities for its production.

236. The *modus operandi* is, to burn the fire down previously until little remains but slag and ashes, push the burnable part to one side, drawing the other out clean. Bring now the burnable part back on to the clean bars with the slice or poker, throw over this some round coal, then draw out the fire clean; then spread the young fire over the bars; fire now lightly until brightened up sufficiently for a charge. The interval between firing will be from 15 to 25 minutes. The furnace ought never to be overcharged with coal. All lumps should be broken to size of a man's hand; and coal once laid on should be allowed time to burn. Where a series of furnaces are worked by one attendant, the order of cleaning should suggest itself, as when one fire has been got into full swing after being cleaned, another may be fixed for *Burning down*, which operation is very simple, viz. put on no more coal, and keep it spread over the bare bars until it is ready for drawing, having always regard to the falling steam.

237. Many expert firemen when *cleaning* one of several fires, prefer to draw everything right out, fill up with green coal, and, putting a couple of

shovelfuls of live coal on the front, leave it to nature. A good hand can have a fire in this way in half the time and without dropping the pressure more than 2 or 3 lb. This suits the fireman better than the boiler, for to have a bright furnace changed in a few minutes into a cold, green, black fire is a barbarous practice. The convenience for the time being cannot compensate for the sudden contraction and tear of that part of the boiler.

238. Long furnaces, as in Cornish and marine boiler fires, are best worked high at front and working them back. Square furnaces, such as locomotives having no bridge, can be worked best hollow in the centre. Firing little and often is the rule for any furnace whatever. It is not good practice to work fires with a green and bright side alternately in bridge furnaces, as it produces down-draught and alternating strains of expansion.

239. The amount of ash, clinker, and refuse tell very plainly whether the treatment or the coal itself is bad. Clinker cannot be avoided in close furnaces as in a house fire. A rapid draught produces clinker more or less with all coal. It has been found good practice to burn clinker over again, by mixing with the coal, the advantage of this being that the clinker keeps the fire open, and gets reduced itself to ashes proper, which also protects the fire-bars.

240. Good firing, good combustion, and economy of fuel are synonymous terms (§§ 19 to 32). These diagrams render further comment unnecessary, showing, as they do, the whole process from the

L

" bing " to combustion in the furnace, and the matter has been practically illustrated, from which it will be obvious that the free admission of fresh air need not be confounded with the overdone combustion, as at § 30. The clear grey and white luminous flame without smoke, characteristic of good combustion, need not be confounded with the dull-red, smoky flame of bad. The difference between underdone and overdone, so important in firing, is like the case of the Scotch farmer who reprimanded his ploughman for not having his furrow up to the straight, while the man held that the furrow was past the straight.

241. On no account throw water into a fire, or ash-pits, or boiler fronts. Keep the firing-place or stoke-hole clear and clean them every watch. Keep all sludge door screws, bridge nuts, or other gear protected from wet muck or ashes, especially when cleaning fires. Have all ashes or muck promptly removed directly it is found. Never keep wet ashes, however small the quantity, against the front of boiler, rather put it into the ash-pit (§§ 269, 270).

242. Cinders that have been quenched with fresh water can be again burned as opportunity offers, but if quenched with salt water are practically unburnable. After cleaning fires, all ashes ought to be promptly removed, and the ash-pits ought to be drawn frequently (§§ 285, &c.).

When furnaces are very large, it is best to have double doors, which protect the fireman from so much exposure to great heat, and offer less opening to the rush of cold, as one door can always be shut.

A hole about 2½ inches square is a great object in the bottom of doors for slicing without opening doors at all; this hole should be provided with a shutter. Fire doors ought to have plenty of air holes; they are easily plugged up if not wanted. Doors need only be large enough for the free admission of a man. The same rule is applicable to the opening over the bridge. For small boilers it may require the removal of a couple of bricks.

243. Smoke

has been long a vexed question, and is simply one of good or bad combustion (§ 30). It is caused by heavy firing, viz. putting too much green coal on a fire at one time, thereby cooling it down and choking out the flame, which is a complete stoppage of combustion, as the carbon in the smoke will not join the oxygen under a temperature of flame heat. Hydrogen in the furnace also produces smoke. The infallible cure is fresh air. To prevent smoke—fresh air; to cure or consume smoke—fresh air. Smoke prevention and smoke burning are, practically, synonymous terms. When two or more fires are leading into one combustion chamber, the method of putting one to utilise or consume the smoke of the other can be carried out most completely. Thus, with two fires, instead of firing both every 20 minutes, fire one every 10 minutes, leaving the door 1½ or 2 inches open; when about to fire the other, shut this door close ; and so on, one door close and one 2 inches open all the time, this admits sufficiency of oxygen ; the bright fire meets these combining quantities of oxygen and carbon with flame absorbing any possible excess of carbon, and accelerates the perfect combustion. Every particle of the carbon vapours or smoke is in this way utilised.

214. Draught

is created by the expansion of gases by heat (§ 30). Rules are observed by boiler builders allowing so many feet fire-grate to so many feet heating surface. Owners of boilers would, however, consult their own interest more by allowing more space for combustion, as also a larger margin against flue areas being choked with soot and muck (§ 149), or in reality larger boilers. The larger the areas producing evaporation the easier and slower the combustion. No room would be lost, as the slower combustion would mean more perfect combustion, which means burning less coal, and that means in turn fewer bunkers, which would then equal the extra space for larger boilers (§ 144).

Draught joins the two subjects of combustion and ventilation. To insure good draught there must be no leakage from other flues or fires entering the same stalk or funnel, unless the fire be kept up in each of them, otherwise, air coming through empty spaces will kill the best draught, which is at that part where gases deliver their best heat, just over the bridge. Beyond that the heat is decreasing all the time to the point of escape.

245. VENTILATION

is a subject which does not appear to be well understood, in proof of which we have only to behold the array of cans, cowls, &c., along the house-tops of any populous place. The remarkable diversities of design side by side prove the existing diversity of opinion. Indeed, throughout, there is evidently no recognised law of ventilation. The promoted ventilation of enclosed furnaces carries us beyond this difficulty so far, but often produces inverse conditions. In many cases the ignorance existing on this subject leads to many boilers being ruined in setting, in so far as economy is concerned.

A. The s.s. ———, of London, had two boilers side by side, a space being on each side of 5 feet between each and the ship's side. One of these spaces or passages was used as a passage way, the other was closed up. The starboard boiler next this passage burned about $\frac{1}{3}$ more coal, and did nearly $\frac{1}{4}$ more work. By-and-by the open one was closed up, and the closed one opened, with this result, that the starboard boiler at once lost its advantage, which the other appropriated. The susceptibility of boilers in this respect is underrated (§ 301).

246. Down-draught

is a buzzing noise caused by down-currents of air meeting the ascending rush of heated gases, resulting from insufficiency of air through furnaces or ash-pits to supply the void caused by the ascending column of heat. It is sometimes a very troublesome condition of circumstances—is, to speak metaphorically, an external priming, or confliction of currents, by furnaces, ventilators, or smoke stalk. The fire will be forced down through the bars, or through the fire doors outwards, similarly as when a damper is suddenly shut. The whole current is the reverse way; the boilers dance as if to a tune. This requires to be at once checked by closing the damper a little and opening fire or tube doors just sufficiently to stop it. It is very remarkable that often very little causes and very little cures this trouble. If unchecked it may, by heating the ashes in the pits, cause the fire bars to drop (§ 250). It is also detrimental to the boiler generally.

247. Down-draught is caused also by fire bars being choked up with green coal. The fresh air rushes down one side of the chimney to feed the fire, while the ascending gases take the other, collision of these currents producing the phenomena in gulps or pulses, as water pouring from a bottle held upside down, or as the gulping of a cask when being emptied from the bung.

The fire will supply itself with fresh air from

the top, if it cannot get supplied through the bars.
Another cause is the bars getting bare at the back,
pigeon holes too much open allowing too much air
to pass up through the combustion chamber, producing such as spiral and zigzag currents.

Irregular firing, badly constructed bridges, ventilators too long or too short, or improperly situated for
the fires, anything tending to disturb the *status quo*,
such as any inequality of combustion or draught
power, owing to opening of fire doors, &c., or any
alteration of the outside surroundings are also
causes of down-draught. Ordinary draught is often
much affected by very remote causes: as little as an

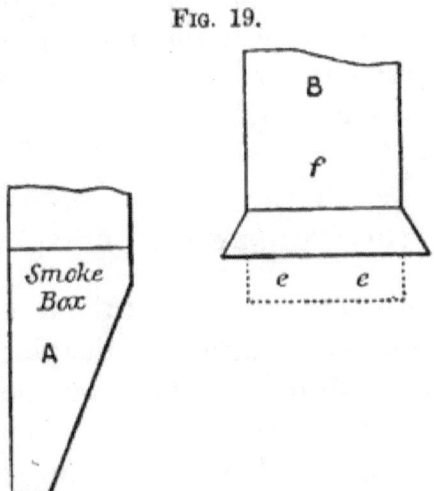

Fig. 19.

empty cask getting displaced is sufficient; in steamers
it is in seven out of every ten cases the fault of
ventilators — a piece of plate fixed inside as at
f, changing the angle of current, a small piece

either added to or taken from the bottom of a ventilator, *e e*, always alters the run of draught in steamers.

248. As a rule the cure is arrived at similarly as with a kindred trouble of domestic experience, down, or back, smoke. The housewife will open the door a good deal and the window a little bit, back door partly shut, the fore one partly open, or smoke board down. By thus alternating these expedients she at last lights on a cure. If, again, the draught is weak, to help a young fire she will stick in the poker with the head resting on the hearth, the passage of the air being promoted by its travelling up the poker into the fire, drawn as if by a lightning conductor. Domestic traditions are more of solid philosophy than the built-up theories which are nowadays brought forward. A kindred trouble is

249. CHIMNEY TOP ABLAZE.

This is an issue of two or more of the conditions producing down-draught, although the phenomenon is different, viz. loose inflammable gases burning from chimney or stalk top with a flame of low intensity, which may be described as a second-rate combustion from the smoke box as a base of flame, or, the bituminous products of the coal are being carried off by the draught unignited, or otherwise not mated with their quantum of air sufficient for combustion, and on their arrival at the funnel or stalk top this deficiency is supplied, as a straggling piece of flame is sufficient to ignite the combination.

Any one can demonstrate this phenomenon to his heart's content with a piece of tube placed over a gas burner as used for heating purposes (see sketch, Fig. 20), a modification of the "Bunsen burner." When the tube is lifted as at A the flame will be at *e e*, but when down as at B it burns at the top, *d d*, just as any other funnel would if getting air down as *f*, which only requires the presence of flame; smelting or other furnaces are all instances of the same phenomena, so is a gas retort. But when the base of flame is from the smoke box, funnel bottom, or furnace itself, then the chimney will suffer; in such cases it is usual to bind stalks with hoops and straps.

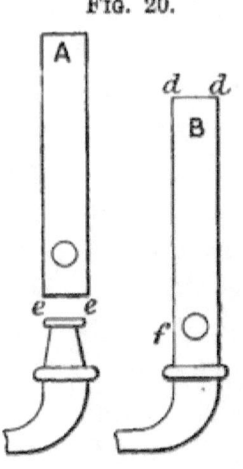

Fig. 20.

250. Fire Bars dropping

is a very awkward matter, and in certain cases a very disquieting one. In seven cases out of every ten it is caused by the ashes accumulating in the pits (Fig. 2, C), preventing the free admission of air through between the bars, which causes the ashes to get hot; the two causes combined heat the bars red-hot, at which heat they cannot sustain their own weight and that of the fire, and, consequently, break asunder by the centre, or bend until sufficiently shortened to drop down by clearing the end. The back ends in some cases are burnt off owing to slag and sand about the bridge stopping the air current. It is also sometimes caused by green coal on back bars, and bright coal in front. Sometimes one will drop from age or flaw, but if two or more drop at one time look out for the whole line of bars.

251. In such cases draw the ash-pits cautiously, ease up the clinker off the bars with the pricking tool. Do not risk the weight of the slice in the fire, and clear where choked; shut pigeon holes if open (pigeon holes, Fig. 9, A B).

252. Lime and sand in coal, as well as neglect, promote the dropping down of bars. It is, therefore, necessary to carry a good stock of spare fire bars. Always allow bars to have a good grip on landing, and always allow for expansion.

Dampers.

253. Of these there are four kinds: one stopping at top, or lid damper; the shutter damper, used generally with boilers set up in brickwork; the well-known butterfly damper, such as are fitted in steamers' funnels, and stove pipes. These all act behind the fire. Another kind is ash-pit dampers placed in front of fires. They act to exclude the air from the fires, the others to paralyse its action. For steamers the butterfly in the funnel is very handy;

but they would be infinitely more effectual if made something like a close fit when shut. It is a mistake that these be fitted with 2 to 5 inches clearance all round, causing ash-pit dampers to be used instead. In working there is no method to compare with that of adjusting four to six ash-pit dampers by the lifting of a single lever.

254. But it is bad practice to work on with two kinds of dampers at the same time. Which ever is the most suitable should be retained; the other should be dismantled right off. Ash-pit dampers can be had very light—No. 16 B.W.G. funnel dampers from 3-16ths to 5-16ths, according to their position for heat. Nothing could be better for land engines than the shutter damper.

255. Damper Gear

should be of the simplest and most direct construction, handy, easy to work, and not liable to catch or jam. The handle or weight when the damper is close should be always down. Very awkward circumstances have occurred through the position of the weight being misunderstood. For such cases a rule is necessary, in view of which, say, we put *down* for shut, and *up* for open, as a handle or weight to be pulled down is suggestive of shutting, and *vice versâ*.

Easing Fires.

256. Close the damper, but never suddenly, to nearly shut; in a few seconds shut close. As a rule, with many marine boilers, the damper shut and tube doors open is the way of checking the fires. It is very bad in principle to open tube doors, although the practice is winked at. Nothing can be expected but leaky tubes where this is in practice. But the question is, what is a man to do when running with heavy fires if unexpectedly brought to?

Sudden Stoppage.

257. In such cases the loss of fresh water by the blowing off of steam is so serious when using salt water as to make all the checking available necessary. Some boilers are fitted with small shutters under smoke boxes; this is equal to opening of tube doors, and very much handier, otherwise it is best to open centre door first. Draw at once half of the fires if heavy, leaving the bars clear at the back, allowing passage of air through set ventilators back to wind. Use off the steam in pumping bilges, or in any other duties in which it can be disposed of. There are also in cases of this kind great differences of different jobs. One manipulation may succeed for checking fires in one, and not in others, but shutting dampers is always effectual.

Keeping Steam Handy.

258. This is often a very unmerciful order for the boilers; it is most in vogue with tug boats and other small vessels. Nothing suits a boiler or attendant better than that the captain, when done with engines for the time being, should state time for steam again, or as near as may be, and then to abide by it as far as possible. Keeping steam handy for the first hour means keeping all fires in check, the second hour putting some coal on fires and keeping the doors open, third hour banking one fire, and so on, keeping steam on the gauge, and coal on the fire that can readily be set away.

Alarums.

259. There are alarums for telling when the steam gets too high or too low, and alarums for telling when the water gets too low or too high in the boiler. Without saying a word upon the merits of these inventions, the fact is all the same, that the use of such things brings about their abuse, which precedes their disuse. It is a mistake to suppose that boilers are safer with such clap-trap than under the vigilance of a proper attendant. When such things exist an attendant depends on them. His own responsibility has been removed, and his vigilance along with it; you cannot now depend upon him, and you cannot call him to account for the fault of these things. You might as well indulge a

sentry with a sofa, wine, and cigars. The water float used by Watt is now obsolete.

A. *Fusible plugs* are still common with the "loco" style of boilers. They should be removed every three months. Their use is productive of leakage and corrosion, while their efficacy has never been proved.

Safety Valves.

260. The leaking of safety valves very rapidly promotes scale, owing to the loss of fresh water. Spring-loaded valves are best, and ought to be turned frequently by the horns to keep them tight. They sometimes leak owing to sand holes in the metal, or grit getting into the seat when blowing off steam. If turning the valve does not make it tight, it should be taken out the first opportunity, overhauled and properly ground in its place, taking care that the joints and screws are all free and in order. There is no excuse for properly got-up spring valves leaking, and nothing more effectually promotes scale.

Fig. 21.

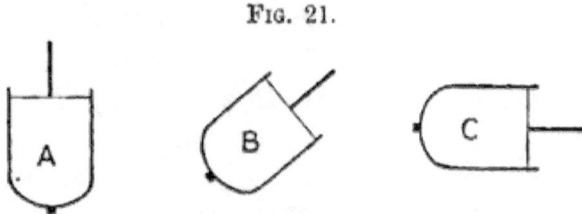

261. Lever or dead-weight valves are not suitable for ocean-going vessels, because, if a vessel carry steam to 90 lb. when upright, as at A, she

will blow off at 45 lb. when heeled over to the angle of 45° as at B. It is then self-evident that if situated as at C, no pressure will be on the valve; therefore, with a ship rolling heavily, and with weighted valves, she can carry very little more than half-pressure.

262. BOILER GAINING WATER

is a sure sign that there is a connection between the sea and condenser, such as supplementary feed cock leaky, or split tube in condenser. Taste the feed water; if it is saltish, that indicates a split tube, which is sufficient to account for a very heavy influx of water. Sometimes with a rolling ship, or discharge valve out of order, water will get from the sea to the hot well. If the surplus water requires much blowing of the boiler, it should be cured at once.

263. If the ship is by the stern, take off the aft condenser door; a leaky tube will show water issuing from it. In such a case, take off the other door and plug the tube at each end. A few tubes plugged up will not sensibly affect the duty of the condenser. The effect of a condenser leaking in this way means an abnormal amount of solids taken into the boiler. The same effect results from a

264. LEAKY BOILER,

and both are similarly matters of anxiety. Marine boilers of the ordinary kind generally leak about the seams or stays in the combustion chambers, or tube ends. It is an easy matter stopping leaks below the level of the fire, as bolts, washers, cement, or any similar stop-gap can be applied. But for instance, dog-stays, or seams, giving way above the line of fire are not so easily treated, as cement in that case cannot be used. We must always find, in such cases, if the part is exposed to the action of the flame. The trouble is often with riveted stays. It is best to bore and tap, say a $\frac{5}{8}$-inch hole, and joint on a washer with a screw pin either above or below water line, which will serve until better can be done for it. Independent of the loss of water, leaks have always a deteriorating effect on boilers. All sources of leakage by blow-cocks, or other cause, should be frequently tested.

HYDROKINATOR.

265. This instrument would be of infinitely more service were the circumstances more in its favour, viz. with a boiler strictly clean and working with clean water. Its mission is that of circulating the cold water underneath the furnaces where the circulation of heat is sluggish (§ 129), and forcing it upwards (Fig. 10). It certainly fulfils its mission in that respect faithfully, but the trouble of its use is, that it circulates all mud and sediment as well, from the bottom up on to the heating surfaces. In other words, salts of lime deposited at bottom innocuous, are forced again into the water, causing a

heavy influx of scale-forming matter on the heating surfaces.

266. Were a false bottom inserted where the action is strong and the circulation by the instrument taking effect off this plate, whereby the bottom deposit would be undisturbed, then the hydrokinator would be a benefit.

VACUUM IN THE BOILER.

267. A very reprehensible practice which exists, especially with tug-boat engineers, is when changing the boiler water to open the blow-off cock, allowing the boiler to empty itself of water and steam, and run up again as it likes, the disposition being to blow water all out, then steam, until the pressure be say 5 or 10 lbs., or proportional to the head of water or sea level.

The cold water now coming in condenses the steam; a vacuum is thereby formed filling the boiler up to the crown, if not shut off in time. As most of these vessels have the water level of their boilers above the sea level, the expedient of filling with the vacuum saves pumping up, hence the practice.

268. All boilers, especially H.P., when new, "weep" or leak more or less when under pressure. However, these "weepings" shortly cease, owing to the filling in of sediment, which tends always in such directions. The sediment thus *trapped* remains, forming a naturally tightened seam as part and parcel of the

M

boiler provided the opening up strain be internal, and by inside pressure. But if we produce a vacuum all the conditions are reversed, the seams open out now as naturally with the vacuum as they were shut up with the pressure. The boiler is worked out and in like a concertina, and leaks like a sieve; which is working its own cure too late.

269. Corrosion

may be put as a species of consumption that iron generally, and the steam boiler in particular, is heir to, induced by its situation and the force of circumstances.

Although, to all appearance, a steam boiler ought to be indestructible, and the last to succumb to disease, yet this is just the reverse of the case. From the moment fires are lit up, the boiler appears to take its place on the list of animated nature and its corresponding share of the troubles and afflictions to be found there. The continual fatigue of alternating strains, volume, and temperature-ravages, as well as its own particular diseases consume it.

Outside corrosion is due to leaks from whatever cause, the escaping water corroding as it goes, and also to dampness or inroads of water. For the former we have to look to safety valves, stop valves, manhole and sludge doors, and the landings of all boiler mountings on the shell, butts and straps, seams and rivets (§§ 300, 301). And for the latter we have, for land boilers, the same leakage, after working mischief on the upper part, finding its way underneath to a settled retreat, where it attacks the shell under cover of the bricks, and all flues, seatings, bridges, and brickwork.

270. The prevention of the former is good workmanship and supervision, as well as design; and for its removal wherever it is found, have the part caulked, scraped, and rubbed over with Stockholm tar (see Pitting, § 133). For the latter, similar treatment is required as well as the frequent removal of bricks for cleaning and inspection. Brick and lime faces should be as narrow as possible, preferably seating bricks that are made for this purpose and properly fitted without the intervention of lime or clay (§§ 300, 301). The corrosion here described is simply the oxidisation of the iron itself, or water combining with the oxygen of the air to consume or rust the iron; the process is familiar to all. This is produced wholesale in marine boilers exposed to bilge water and covered up with old fire-bars, bricks, &c., which being a visible operation has no excuse; and in land boilers planted in a low situation subject to inundation and continual damp, as also to the ravages of rainfall.

We have also corrosion in the furnaces, combustion chambers, and flues from combinations of sulphur and other acids (§ 285); also inside the boiler from grooving (§ 313) and Pitting (§ 133).

270. Rusty Boilers

are very common. The difference between this and pitting, grooving, &c., is that this is general, while the others are local. New boilers are often subject to it. The cause of rust in steam boilers is the same with that of the new washhouse boiler or new pot; the cure is known to all frugal housewives, and is simply washing soda (§ 185). It is caused either by acids in the water or other substances put into the boiler, or by the natural oxidisation of the iron owing to its being allowed to remain wet (§ 133, A, B). With steam boilers when this rusting is very severe it is rather serious, and requires prompt attention; two or three doses of caustic soda, and then the occasional use of common soda will cure the trouble, but in all such cases the boiler must be properly scraped and cleaned. This is a more rational cure than artificial scaling (§ 102).

Chipping Heating Surfaces.

271. The formation of scale in steam boilers, although a necessary condition from the use of dense waters, entails much labour and expense for removal, regarding the necessity of which see §§ 192–195.

Owing to the running of many steamers it is next to impossible to have an opportunity, with boilers properly cooled, to allow of systematic scaling. To accomplish this at all, we require to scale one boiler one trip, and another the next, or, if hard-pressed, one box of tubes each trip. But some method or course of action must be fixed, otherwise labour will be thrown away.

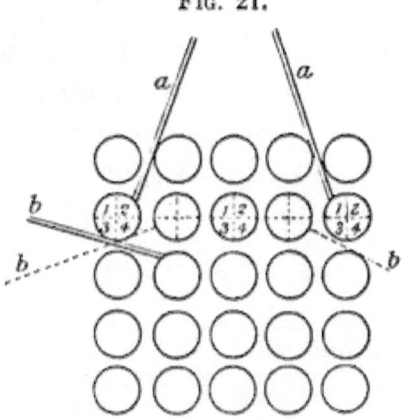

Fig. 21.

Let us then select, say, the centre box of tubes, and for expedition we put on two men, as at *a a*, slicing downwards. This will scale 1 and 2, or top half of tubes. A man at each side *b* and *b b* will scale 3 and 4 or under half. Slicing downwards, as at *a a*, is often much interrupted by stays, in that case all the scaling can be done from the sides, *b b* doing 1 and 3, and *b* 2 and 4.

Commencing at top 1 we remove a quarter side on each top side downwards, as in diagram, but as slicing upwards is more difficult, it is best to slice as far as possible from the top. We can do one side of all the tubes right down first, then the other sides, finishing up by slicing from the sides what may be remaining on the under sides of tubes. This done, we have free circulation from furnace crowns to water surface.

272. In view of its great importance this vertical

or down slicing should invariably be accomplished first. In slicing across it is advisable to have two men working opposite to each other, they being a check on each other. For speed, four men can work on a box of tubes thus:—One man begins at X and another at W, both working in the same direction, and never fouling each other, both working to the right.

Fig. 22.

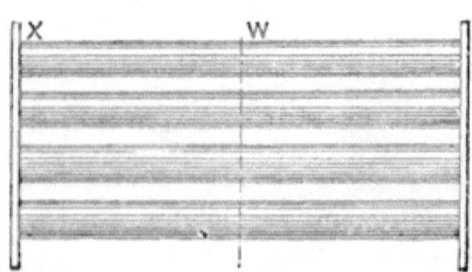

By this method the engineer can see at a glance the progress being made, and eventually get his job accomplished. Otherwise, sending in lads or men to work indiscriminately, each taking his own favourite part, with no check upon them as to whether they play or labour, is throwing money away.

273. The heating surfaces are subjoined here for scaling in the order of their importance.

For Safety.

1. Combustion chamber, crowns, and Galloway tubes.
2. Furnace crowns.

3. Combustion chamber backs.

4. Combustion chamber, tube-plate, and tube-necks.

5. Smoke-box tube-plate.

6. Combustion chamber sides.

7. Tubes.

For Economy.

7, 3, 4, 2, 1, 6, 5.

Scale on the outside shell has the virtue of preventing the radiation of heat, and is therefore better to be left on.

BANKING FIRES OR DAMPING.

274. To do this the fire must be well burned down, then pushed back against the bridge, clinker and all, according to circumstances; cover all up with wet ashes and wet slack coal, leave all fire-tube doors and dampers open. If the fire kindle, shut the damper and cover bright parts of fire as it breaks through. If the fire threatens to go out, shut the tube and fire doors. This is termed banking fires back. The other way of banking fires to front allows the boiler to cool more, and is not so apt to break through in flame as those banked at back. For banking where steam might be wanted in a hurry, bank at back; if not, bank to front.

"HOLING" THE FIRES.

275. That is just a phrase used to mean half-banking the fires. The back end of furnaces are

scraped forward, leaving the bars bare at back. This answers for keeping steam handy.

TUBES BURSTING AT SEA.

276. This is often a very unpleasant business. Tubes burst at sea from the following causes:—1st. Badly welded seams tending to open. 2nd. Pitting (§ 133). This occurs generally as a small hole on the under side of the tube, and on the side next the water. 3rd. Leaking of tube ends.

This last is the common trouble with salted boilers and is the result of the necks of tubes being so covered with heavy scale that the water is incapable of cooling the part. The flame acting severely on the plate on the one side, and the water being cut off from cooling on the other, the side next the flame will of course expand with the heat. This expansion on one side produces a bag or *buckle* outwards towards the heat, thus disengaging itself from the tube ends—in other words, the plate leaves the tubes, and heavy leakage often follows. Very often the buckling cracks off the scale, allowing the water to get to the part, which will prevent further buckling of the plate (which is known also by the term "burning of the plate"). We may temporarily stem the leakage by expanding the tubes with the expander; but the only cure is the removal of the scale, although we can never take the buckles out of the plate.

277. A desperate cure may be had by driving the tubes from the smoke-box through, say, $\frac{1}{8}$ of an inch, which will of course start the scale about the necks

of tubes, which will make its removal easy. This artifice cannot be often resorted to with impunity.

278. Sometimes tubes are defective from the first, sometimes injured in putting in, and injured in scaling, but generally tubes do not give the same trouble now as thirty years ago, being of better manufacture, mostly solid drawn, and the holes in the tube-plates being carefully sized to suit the small difference of diameter, the smoke-box end being about $\frac{1}{16}$ larger; and when being driven home the job is metal and metal before the expander is applied.

Leaky Tubes.

279. A leaky tube may be detected by a humming or hissing noise within the smoke-box. If the leak be near the smoke-box end, water and steam will issue from the bad tube, but if about the centre or towards the other end, it can be detected in this way:—On opening the fire-doors the water will be seen falling about the bridge or escaping on to the face of the chamber. If so, open a tube-door, and find the row of tubes where the leak is heaviest. The bad tube will very likely be the uppermost of those that are dropping water. A handful of coal-dust thrown over the fire from time to time will cause a flame of light at the back, which will show the leaky tube. Failing this, with your hand and a piece of canvas, stop the tubes individually about the part. When the noise ceases or is changed your hand is on the bad tube. Tubes always show the water on the after side owing to the deeper draft aft.

Tube Stopping.

280. The old method of stopping leaking tubes with wooden stoppers is not much in practice now, being superseded by handier methods. But as wooden plugs can be got when patent stoppers cannot, the rehearsal of the *modus operandi* is a matter of prudence. To be acquainted with any primitive or simple adaptation of means to an end is a stern necessity with all who go down to the sea in ships.

Fig. 23. T T, tube-plate; *e e*, leaky tube; *l*, situation of leak; *p p*, wooden plug of any soft wood; *c*, the bridge; *d*, the furnace; *g g*, showing difference of level between the two ends of tubes.

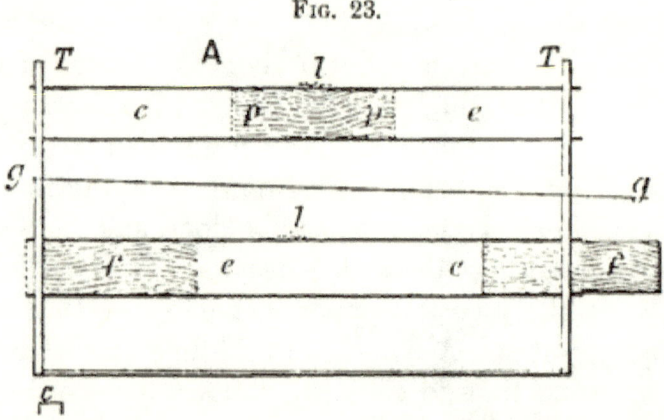

Fig. 23.

The plugging requires to be performed as at *p p*, and the plug is driven a very good fit, but not so tight as to strain the tube. If we have dense water in the boiler the stopping will be facilitated by salting. Plugging never stands done at ends as at *f f*, being, in the first place, burned by the fire, and then shrunk with the heat. This is only available for a short run, and although the plugs are fitted tight they will not stand pressure over 35 lbs. to be depended on, beyond which they lose grip and fly. The effectual plug-stop is when planted as at *p p* against the leak *l*. If perfectly dry when driven in it will immediately swell in its place, making a complete stoppage, with no chance of getting blown out. If near the chamber end, it may be best to draw a fire and plug it from that end. If the leak is heavy, reduce the pressure until the plugging is accomplished.

281. The patent stoppers with rubber discs are very handy and effective if the proper size for the tubes, but if too small are useless, and if too large are very treacherous, getting fast in the tube before arriving at the proper place, when it often is a hard job to get them out. To insure success, the length of pipe must be equal to the distance between the tube-plates, minus the thickness of the two iron washers next to it (Fig. 24, B). The two washers at the back end ought to be $\frac{1}{8}$ less than the inside diameter of tube, and the washers for the smoke-box end $\frac{1}{16}$ smaller than the inside of the tube; the rubber discs from $\frac{3}{8}$ to $\frac{5}{8}$ thick, and just a little less than their respective washers. The stopper must be put through with the nut slack. Sometimes much trouble arises from the stopper being pushed too far through so as to drop, which is a bad job, a man having to get over the bridge to replace it.

282. To entirely obviate this it is good practice to have a large washer next to the front nut, and about 5 or 6 inches in diameter, as at *n n*. The stopper can now be pushed straight home into its place with certainty, when a few cants of the nut *l* will expand the

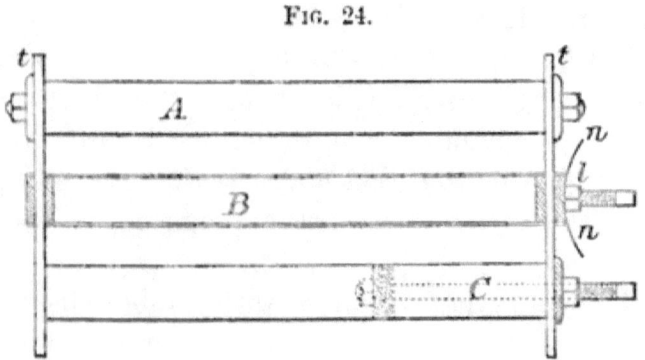

Fig. 24.

rubber discs at both ends simultaneously. The tube is now stopped. This large washer *n n* also serves the purpose of protecting the operator completely from the rush of scalding water and steam from the tube. Care must be taken that these rubber discs expand exactly at the tube plate *t t*.

283. It is the practice of some engineers, with a tube leaking near the front, to put in a short patent stopper, as shown in Fig. 24 at C, which is very bad practice if the tube is disposed to split further as the rubber disc expands, which, owing to its not having the encircling support of the tube-plate, it is very apt to do. Such stoppers ought always to be used full length, otherwise a simple matter may be made a bad job.

284. The most effectual of tube-stoppers are the well-known bolted flanges (Fig. 24, A). These are sometimes made hollow to receive the tube ends, these projections being left on, which is best when the leakage is by the inside of the tube. If the tube ends are fair and square, any jointing of gauze wire, insertion, &c., will, if not too much exposed to violent heat, stand well. When the leakage is by the seam, and cannot be stopped by the expander, we must cut the tube ends off flush. If the part shows a good plain face, the joint may be insertion, canvas, or gauze wire; but if otherwise, a light grummet of three-plait spun yarn, each strand having two threads, or what is better, a well-made scarfed and twine-wound grummet of ½-inch Tuck's packing. When well done this lasts for months, although liable to leak when raising steam. But it must be borne in mind that tube stopping is only a stop-gap, and that no stopping is so satisfactory as the tube itself, loss of heating surface aside. For high pressures wood stoppers are only good from a few hours to a few days when done as at pp, Fig. 23. They will often go for weeks under favourable circumstances—that is, the plug put in very well seasoned, and the water dense. Of course, freshening the water starts the leak anew. Stopping, as at ff, ought never to be practised at all. Nothing can beat the rubber disc stopper (Fig. 24, B) for stopping bad tubes at sea without letting down the steam pressure. They can be afterwards disengaged from the boiler and superseded by tubes. With new rubber washers they are again fit for use.

Concentrated Flame.

285. The action of the flame when concentrated in the combustion chamber is often injurious to such projections as stoppers, tube ends, stay-nuts, rivet-heads, &c. This occurs with perfectly clean boilers as well as with dirty, when under a certain set of circumstances, such as at §§ 150, 333. The whole flame energy will be concentrated at one spot, and the action of hydrochloric, sulphuric, and other acids in the coals when they join issue show very plainly, by the ravages made, that with heating surfaces there must be a limit to the thickness of the plate. The interposition of soot and muck serves to protect the parts thus exposed from the attacks of coal acids as well as flame. But a slight alteration of the bridge x will stop these attacks.

286. The same phenomena here described give also results inside the boiler.

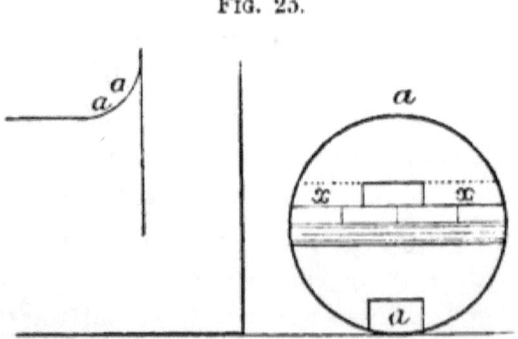

Fig. 23.

There is very often a scale of great hardness and adhesiveness at the junction of the furnace tube with the combustion chamber tube-plate aa, which is the most intractable scale found in any part of a boiler, and should always be promptly removed. It is generally a patch of from $1\frac{1}{2}$ to $1\frac{3}{4}$ foot square in area, resulting from the intensity of the heat of the concentrated flame over the bridges jointly with the small volume of water backing up for cooling two faces of plate as a corner, or owing to scale falling down off the tube-plate.

BRIDGES.

287. Very little alteration of a bridge alters the run of the flame. Cases are known where 10 per cent. of fuel was saved with a Cornish boiler by simply altering the position of one brick on each side of the bridge x, throwing the otherwise escaping flame on to the plate.

288. A good deal of economy can be secured in the construction of bridges whereby to receive the concentrated effect of the flame. The run of flame is easily observed when the fire is under weigh and suitable. Loose bricks can be laid on top and manipulated until the desired run of flame is got—that is, the more the flame is made to impinge on the plate during its escape the more is economy promoted. Regard must always be had to the area of opening, but all the alteration wanted does not amount to much. The area over the bridge should be one-third that of the furnace.

289. When the flame is so concentrated as to be

injurious, it is easily divided, and thus spread over a larger area.

When tube ends and such like are burnt, along with muck getting on to the combustion chamber back, then the bridge is too high. When we have coals and ashes over the back in quantity, it is too low.

GALLOWAY TUBES.

290. The injurious effects of concentrated flame give but a small percentage of risk compared with the general economy. Both sides of the question are well demonstrated in the case of Galloway tubes being placed to receive heat that would otherwise be lost;

FIG. 26.

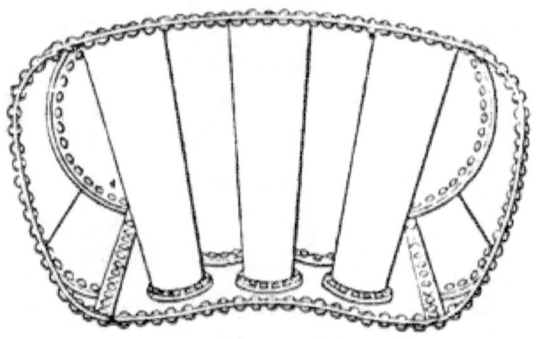

as where the flame charges over the ordinary boiler bridges in the direction of the back plate, but is drawn off by the draught towards the tubes before reaching the plate, and then not being long enough to reach the tubes, is therefore lost, there being no contact with the gases for ignition. This evident loss of heat is obviated considerably by the use of Galloway tubes.

291. Galloway tubes take front rank in the heating economy of boilers. The principles involved in this particular feature of economy have little place in the minds of boiler builders. First cost, and perhaps their susceptibility to overheating, cause their economy as a heating surface, for the present at least, to be ignored. They are most effective when horizontal, or across the flame. Being thin, they convey the heat quickly to the water. In certain circumstances it is difficult to have them stand without the joints leaking; this is when they are bound endwise. With the flange necks constructed as in the sketch they stand very well, but are most successful situated as in a combustion-chamber, where they have facility for expanding; they require care, from the difficulty of keeping them free of deposit. These tubes, situated in the combustion-chamber, are very effective, receiving the concentrated heat of the tongues of flame that would otherwise escape. Their distance from the bridge or each other is an important feature in their economy. The flame, after striking one tube as per sketch (Fig. 27), is just allowed to recover the shock and reconcentrate when it is again dashed against another surface, like a line of waterfalls.

292. The water first takes the bucket, its weight being the force until it arrives at the bottom of the wheel, where it runs nearly level a short distance, repeating the process over and over, until it at last reaches the sea. The same principle extends to the working of steam boilers, in the using of the steam

at a high pressure in one cylinder, then at a low in another, until it is worked out. We get our work out of the water and steam by hanging on to them during their period of escape. The question is, if the same cannot be carried out with the coal.

CONCENTRATION OF HEAT.

293. The most practical application of this idea, and one wrought similarly in stages, is to place a heat receiver at a point proved by trials to receive the heat in greatest effect, which we will call the focal distance—just as when using a concave lens or sun-glass we find the focal distance by trials. The whole heat passing through the disc of glass will be concentrated to, say, $\frac{1}{20}$ at the focussing point. Consequently at that point the velocity of the heat will be increased in proportion; therefore, concentration means also velocity, which is regulated by the pitch or focal distance, draught, and kind of coal, just as the focal distance of the lens varies with its form. This can be demonstrated also with the blow-pipe; the harder we blow the pipe, the farther will the heating distance be removed, and the heating focus can be thus shortened or lengthened at pleasure.

294. So then is the flame over the bridge with increased or decreased draught. Therefore with the varying conditions of draught with coal fires, no hard and fast lines can be laid down. To obviate this drawback, we just abide by a mean focal pitch for the position of bridge.

295. In accordance with this is the position as laid down at Fig. 27, A, so also are the tubes in the flue, as at Fig. 26, at a mean focal distance from the root of flame, thus receiving the heat in flat contact at a given velocity and penetration. The distance between A and B allows the flame to recover the shock, reconcentrate, and dash again against C, as

Fig. 27.

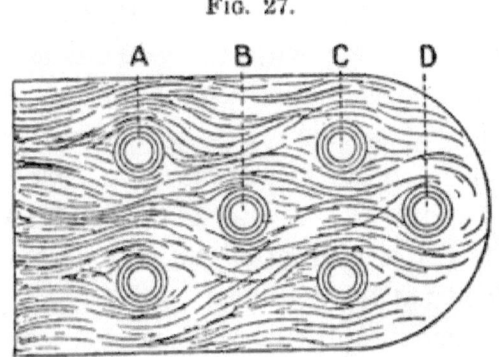

before, the focal distance shortening each charge, until at length the flame, shorn of its heat, expires, its remains going up the chimney.

296. The thinness of plate used for such tubes consequent upon their smallness of diameter, constitutes their great virtue in heating. Their liability, when placed horizontally, to gather deposit can be reduced to a minimum, by giving them plenty of slope, to allow of deposited solids working out. Their tendency to overheating is a proof of their efficacy. To deal with flame when highly concentrated is similar to dealing with pressure or any other concentrated force. We must guard against

results such as scale by overheating of plates, as also other results more painfully disastrous. If concentration be overdone, as at § 333, and accident result, such would not of course favour the method, although proving its efficacy.

CONCENTRATION OF FLAME.

297. The heating effect, as with the bare walls Fig. 28, C, compared with the corrugated walls, as at A, will, firstly, be in ratio with the respective areas of plate, to which is to be added the effect of the concentrated flame striking the corrugation at half-flat contact, also the interruption detaining the heat for its better extraction without destroying the mean velocity of the draught.

If a corrugated furnace is to be, say, 6 feet long, then this 6 feet of furnace when straight would

FIG. 28.

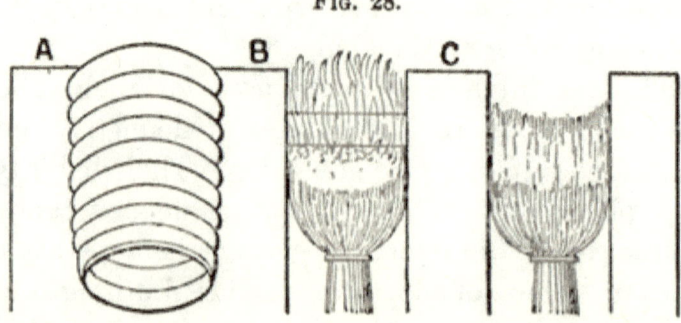

be 8 feet of plate. As on a 6 feet space an 8 feet medium is got for distributing the heat, as a coil of pipe or as a worm distributes the *cold*, just similarly do these coiled furnaces facilitate the distribution and development of heat.

The effect of a flame in a tube or common plain funnel, as compared with interrupted flame, as A and B, is, that the flame enclosed in the tube may be said not to act on the plate at all. The edges being condensed by the cold contact, it passes insulated, as in the common stove (Fig. 12). The cross tube will, if properly focussed, receive the full benefit of the flame. The corrugations have another virtue, that of stiffening the tube, each corrugation acting as a hoop.

UPRIGHT BOILERS.

298. Upright boilers are the least susceptible to scale of any; often known to be working at great density with no appearance of scale; this is owing to their inherent property of circulation, which is best promoted in these boilers. The steam as it is generated has a free margin of escape or circulation to the water surface. Owing to the taper of the heating shell, the path of circulation is defined and direct. This free circulation keeps the heating sides clean. The crown is kept free of scale from its being so near the surface of the water, which is always pure and fresh (§ 35).

299. An engineer was sent on board a steam lighter to examine the boiler — an upright of 12 horse-power, with three Galloway tubes—under the charge of a driver. The enquiry was owing to the

vessel having deteriorated so much in speed. The engineer found that the driver had got an idea that changing the water frequently was his great look-out, and for this purpose he always used the scum-cock, keeping the water under $1\frac{1}{4}$ density. This man, like many more, believed that the water in a boiler was of equal density, top or bottom; also that the deeper his salinometer sank in his boiler water, the less he would have of scale. These mistaken notions ruined the boiler, which was now leaking heavily from Galloway tubes being burnt owing to salt. He did not think deposit in these tubes was important. The boiler was sent on shore for new Galloway tubes and overhaul.

It is a notorious fact that nearly all boilers not fitted with a testing or salinometer cock have their tests taken from the gauge-cocks, in total ignorance of the absurdity of the practice (§ 35).

LAND BOILERS.

Egg-end.

300. These outside-fired boilers are the first to be injured by mud or deposit in the bottom, and the last by shortness of water. They are now common enough flat-ended and stayed in the centre. Their construction is very simple, as they are only a shell, and the cheapest of any. They have been condemned wholesale by authorities on boilers, on account of unfitness to work impregnated waters. Notwithstanding, they have had a long share of

public favour. If these boilers are provided with clean water and a careful attendant they do fairly well, as they are easily kept clean and in repair. They are best set with a declivity towards the back

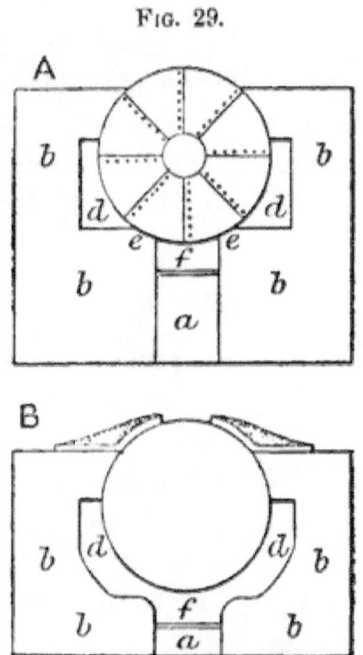

Fig. 29.

end of half an inch to the foot, getting the feed at the front and having the blow-off cock at the back properly protected by brickwork. With this boiler the heating surface is hollow instead of round, which seems to facilitate the cracking of the scale. This gratuitous benefit is owing to the rapid cooling when fire-doors are open, and consequent contraction of plate, which bursts the scale as it forms. They are not adapted for sea. Being well raised off the ground, they do not suffer from damp.

The common setting is as shown at A, Fig. 29, with boiler resting on two points, ee; f, furnace; dd, flues returning and going, the chimney being at opposite end.

The setting as at B, Fig. 29, where the boiler is hung by four riveted brackets, is the easiest and best. The conditions for combustion are more favourable; there is no opportunity for leakage and consequent corrosion, as at ee, Fig. A; bb, brickwork; a, ash-pit.

Cornish.

301. This has been the pioneer of internal heating boilers, pioneer of high pressures, and inside complication.

Fig. 30, A, shows a Cornish boiler set on centre ridge, with flue returning and going, as at Fig. 29, A; f is the furnace; a, ash-pit; dd, flues; bb, brickwork. In this setting the damp or leakage does not effect so much lodgment at its landing as at B, which is a double-flued Cornish boiler, termed in the Midland Counties the Lancashire boiler. It is often not difficult to drive a sounding hammer clean through at the settings owing to corrosion, as at Fig. 29, A, and Fig. 30, B.

The sounding hammer is only proper in the hands of a tradesman accustomed to such work. Take a good-sized chipping hammer, not less than $2\frac{1}{2}$ lb. weight, and having all well cleaned, ply the hammer on to a part you know to be sound and in all respects similarly situated to the parts you suspect; not tapping, but a good firm blow; if you

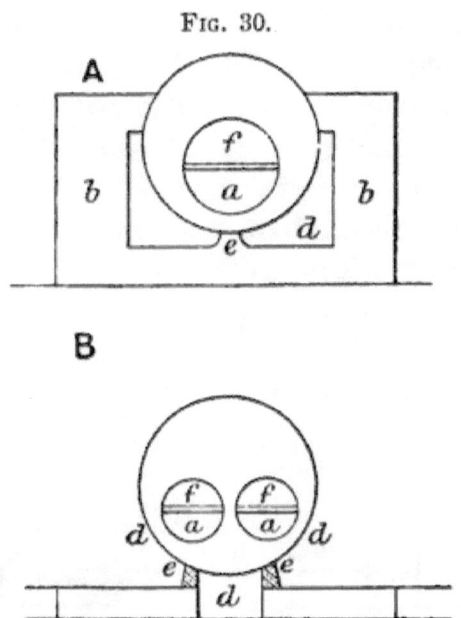

Fig. 30.

send the hammer through the plate so much the better, as you have found out the bad spot. Always compare the sounding of the part you suspect with what you know to be sound. On finding a part with a sound unduly hollow, bore a small hole through it if practicable; after sizing, fill up the hole by a tapped pin or rivet.

Boiler Repairs.

302. Boiler repairs ought to be done by boilermakers, having the necessary skill and tools at their

disposal. Boilers at sea, and in remote situations, however, have often to be temporarily repaired by those in charge. The bulk of such repairs are external, nine-tenths being inside the furnaces or chambers, such as tubes leaking, stay ends leaking, plate-landings leaking, &c. If a rivet leaks, see if it is broken on the inside; if so, at once knock it out, and put in a good fitting bolt filling the hole well, but do not drive it tight. Amateur boilermakers must never attempt to put in rivets.

303. Where a leak occurs from a seam it should be stopped by caulking, making sure that it is in a proper state for it, because too much caulking will make a seam leak worse. Fig. 31, A, shows a case of

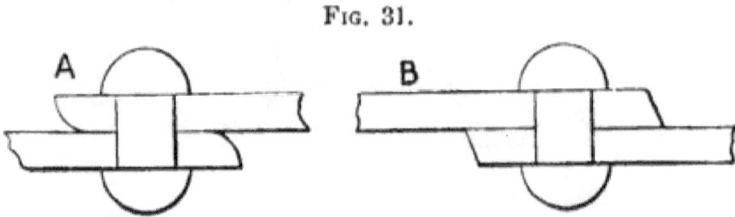

Fig. 31.

over caulking, and can only properly be recaulked when the edges are dressed by clipping, as at B. The question here comes in—will the lap stand it? that is, will the plate when dressed such as B have sufficient margin of strength from edge to side of rivet? which ought never to be less than that of the diameter of rivet.

Caulking should be attended to in all cases. If not directly a matter of safety, it is a matter involving corrosion and deterioration.

304. The engineer of the s.s. "——" was requested by the master of a steam tender to look at the boiler, as it was leaking badly after being just repaired. This was an upright boiler. Some time previously a split in the plate over the line of fire was covered with a patch, the space being cut out, and patch partially checked—in every respect a good job, only it was not large enough in one direction, the split having now extended beyond it. Other parties knowing less about the business executed this second repair. They put on a plate uniform in size and position with the other, but bolted. One hour after the vessel had left port this bolted patch buckled inwards like a great bag, and leaked like a sieve, the driver being unable to keep water in the glass. When the engineer examined it the whole patch was red hot, and leaking at every bolt, indeed within an ace of explosion. The fire was drawn, patch taken off and examined. The bad part had not been cut out of the plate, the holes had not been properly marked off, and were a full sixteenth larger than the bolts, which were too large of themselves, and had enormous grummets under the heads and nuts. The patch itself was jointed with canvas and red-lead; no fitting of the plate or caulking had been attempted. A more complete bad job could not be conceived. A new boiler had to be got. The parties that did this repair were bridge builders.

305. No one should attempt boiler repairs who does not understand it. A boiler should never be struck with a hammer with pressure on, such as at man-hole

doors and such like, always reduce the pressure before doing so.

Holes in patches should always be bored, burr edges taken off, and provision made for caulking. The thinner the patch, the better and easier the job. The bad piece should be cut out. Never tighten up bolts within an ace of breaking, as when the strain proper is added this may prove *the last straw that breaks the camel's back*. This is how bolts are broken.

Riveting	..	..	..	100 = strength of plate.
"	..	..	..	98 = double riveting.
"	..	..	..	76 = single "

Salt Deposition.

306. The donkeyman of the s.s. "D——" reported the water in the gauge-glass being queer, which was the case, and very difficult it was to point out a water line. The fire was drawn and steam blown off; the water was run out from the sludge door as it could not be blown. The furnace top was covered to 2 inches thick with a segregated white mud. The Galloway tubes were about ⅓ full of this same substance, as well as the bottom, at back, opposite the fire-door, to about one-fifth of the circumference of a similar but darker substance. The inside shell was bagged towards the fire as at $a\,a$, $a\,a$, Fig. 32, and full of solid matter, which could only be removed with great difficulty. After much trouble two lumps about the size of cocoa-nuts were detached; in section they were as the sketch. Next the fire shell was a thin black scale like flint. The part $b\,b$ was a com-

pact mass of salt, pure and fit for table use. This mass was put into a bucket of fresh water; eight hours after the salt part was melted out, leaving a hollow shell and the water salt in proportion.

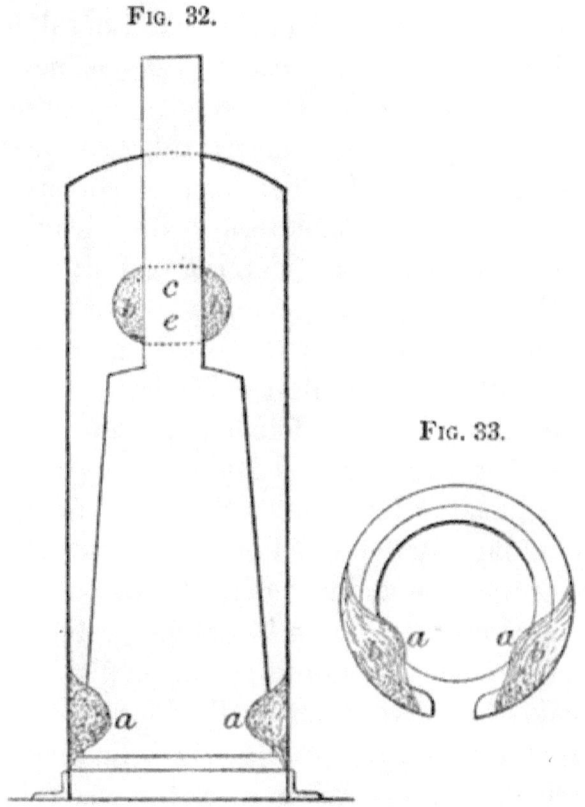

Fig. 32.

Fig. 33.

307. Blow-cocks, gauge-cocks, test-cock, scum-pan, every orifice was choked hard with salt, not lime, but common salt.

The mass $b\,c\,b$, Fig. 32, was a most extraordinary deposit of crust like a white pillow enveloping the funnel at water space, and when removed was a very plainly stratified secretion of lime scale and common salt, corresponding as minutely as possible with the filling up and ebbing of water from time to time and the acts of firing. For the most part it consisted of a common scale about $\frac{1}{8}$ inch thick, like sulphate of lime; the outside crust of each layer was more like common scale, the inside or first coating of the layer being more crystalline. The third layer from the iron was about $\frac{3}{8}$ inch thick and of common salt. One-half of this mass came off bodily, leaving the plate clean, but burnt and deteriorated, c; the other part still adhering, being forcibly removed, a good part stuck tenaciously to the plate, as at $c\,e$; this was also salt, and a similar substance, presumably nitrate of lime. It was found impossible to define the line where the scale began, the iron and scale being so blended into each other's substance; the iron being largely deposited in flakes of oxide. The first layer was iron and salt mixed, $\frac{3}{8}$ inch thick, then 1 inch of common salt, the layers now being similar to the other side. Examining the iron plate, a heavy hard scale of iron was found still adhering. This on being removed left the plate exactly $\frac{3}{16}$ inch thick, the original thickness being $\frac{5}{16}$ inch. The whole of this internal part of the funnel looked as if completely burnt.

308. One counterbalancing advantage is the great density at which these upright boilers can run—often as high as $\frac{6}{33}$ inch, which fact is a corner stone for the non-blowing school. The circulating superi-

ority of this kind of boiler is, however, too much depended on, often with similar results to the above. Salt deposit is not a visible growth, but comes like a thief in the night. It is not uncommon for a vessel having two good donkey boilers, to have them both *hors de combat* in a short time from the same cause.

309. The s.s. "D——" left the port of H—— for G——, being a run of nine days. The tubes had a heavy scale, and during the passage out they leaked badly at ends; much feed water was also lost, owing to feed pumps being in bad repair. The water had been in the boilers three weeks, being once partially changed. The chief engineer took the management of the boilers on his own shoulders, as to scumming, blowing, and feeding, and kept the salinometer in his own posession. Half-way on the return trip one furnace collapsed, as per sketch, Fig. 34, the plate along the line of collapse being marked in strong blue and yellow patches; the other furnace had the same colouring, though not so strongly marked. The chief, who belonged to the non-blowing school, changed his creed for the nonce, and blowing and pumping up was the order of the day. After some time the pressure was reduced. This collapse extended to about two-thirds of the length of furnace; and being thus bagged inwards, the end was strained accordingly from dotted line ee, the forward seam with the shell at the top being sprung and leaking badly; as the pressure was reduced the leak slackened. This boiler was blown out and refilled, and for several days freshened at frequent intervals. On arrival in

port the furnaces had uniformly about $\frac{1}{8}$ or less of scale, the parts collapsed being nearly the same, excepting one or two patches which were darker in colour.

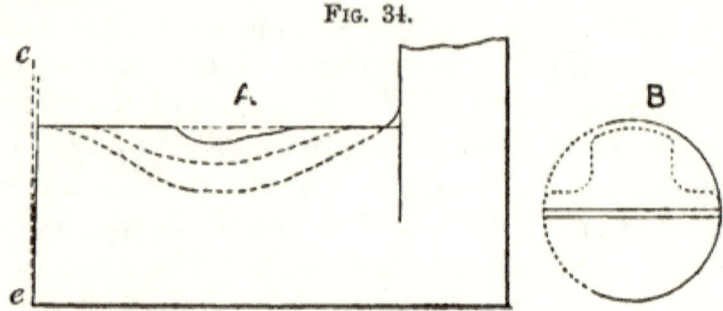

Fig. 34.

310. The tubes in both boilers had $\frac{7}{16}$ inch scale, and leaked badly all the time, although twice expanded. Tube-plates and combustion-chamber had $\frac{5}{16}$ to $\frac{7}{16}$ inch scale, with tube-plates badly buckled. Chamber tops had $\frac{1}{8}$ to $\frac{1}{16}$ inch scale. This last blowing down in port being hurriedly done, added $\frac{1}{16}$ inch to the scale all over.

The scale of a bare eighth was not enough to account for the collapsing, and at no time had the water been out of the gauge-glass. How then is this collapsing to be accounted for? In the first place it may reasonably be put down that the water could not get into that shape without being red hot.

311. The job was stiff for steam, owing to heavy scale on heating surfaces. It was leaking badly by tube ends. The feed-pumps were in bad order. Burning Welsh smokeless coal—no flame—high bridges. Water very dense. No better set of con-

ditions for collapsing the furnaces could exist at the same time. The job altogether was upon its last legs from neglect. The solution is simple enough. The boiler was perfectly choked up with common lime scale, producing heavy leakage, the result of neglect of, among other things, the feed pumps; so much water being drawn from the sea along with its solids into the boiler. One particular of collapsing when the crushing pressure is uniform, is, that the flue is drawn in on both sides of its axis, as at Fig. 34, B.

312. TEAR AND WEAR,

as applied to the steam boiler, is simply *tear*, *wear* being nearly nil. The ravages of wear can be calculated and counteracted by simple devices; but the ravages of tear, which means simply the increase or decrease of bulk, owing to the increase and decrease of temperature, are inevitable. Were it otherwise, steam boilers would last for generations. This tear, from a middle standpoint, is a twofold force, viz. expansion and contraction. Although the one is just the inverse of the other, yet, generally speaking, the expanding is more manageable than the other, just as a tree's roots, in growing, expand as they best can without regard to straight lines. But contraction is uncompromising and irresistible, adhering rigidly to the straight lines, no yielding, bending, or squeezing, but "crack!" and "snap!"

313. Grooving,

when it occurs, is a serious matter, as if it be extensive, the boiler is doomed. The origin of grooving is mechanical, and is due to alternating strains existing at certain parts of a boiler while at work. When once excited, the action rapidly developes, becoming chemical. Its operations are most active after the boiler has been emptied of its water. Locomotives, tramcar engines, travelling crane engines, &c., and all boilers exposed to jolting, or tear and wear from without, are liable to this. The mechanical action firstly strains and fatigues the plate until the skin is fretted and destroyed, it then becomes an easy prey to any acids in the water, or the action of the water itself. It is to be looked for about any shelved or ledging parts, such as landings of flanges, gusset stays, tops of lap joints and flanges, angle iron ends, or, particularly, where a stiff and weak part join, as where a piece of strong new plate may have been used to patch an old boiler.

314. The exciting cause with fixed and marine boilers is the expansion and contraction. The exciting cause with unfixed and locomotive boilers is similar to that described, viz. where the rigidity of one part opposes the breathing or pulsating of another part, added to the fatigue of jolting and knocking about.

If we take a piece of iron wire and bend it backwards and forwards as if we wished to part it, the skin will fret and break, the fibres of the iron open up, and its substance, thus chafed, is gradually being detached; this is just the condition favourable to the attacks of water.

315. Every one knows that iron filings put into a vial with water, will become in a short time a strong acid, attacking iron savagely. It is also well known that in boilers liable to grooving, there are more or less facilities for the process, viz. these nooks and corners where there is always a little water left, or, being so often left, it may be very well termed an acid (§ 181). If a fretted spot is started during working, it resolves itself into a big globular drop when cold, bursts, and runs where it can discover a fretted part of the skin of the plate. The next heating up will repeat the process, this acidulated water following exactly the previous run. We then very soon have a groove which will ripen into a cankered open rut, and sometimes a chain of ruts and holes, making it hard to draw the line between grooving and pitting.

Pressure Gauge.

316. This is too well known to require description, yet it is safe to say that not more than nine-tenths of those who are familiar with, and have to use it, understand it. The question, "Does the pressure gauge tell the total pressure on a piston, or the pressure above the atmosphere?" will be answered correctly by few indeed. It tells the actual

steam pressure measured above the atmosphere. Thus (§ 221):—

Steam pressure, say	30
Vacuum ,, ,,	15
Total pressure	45

Of course, with all high-pressure engines there is no vacuum. The syphon pipe is a bend to hold water to act between the steam and the gauge. This water should not be blown out at all, as the steam, acting direct on the instrument, soon renders it useless. They would serve longer and better if not provided with pet-cocks at bottom of bend at all, as most boiler attendants work on with these cocks blowing and spoiling all the while, as they know no better. The pressure gauge must be placed in a safe and prominent place, and the steam tube led to it from the boiler crown or steam dome. It should never be interfered with further than occasionally giving it a slight tap with the hand to excite it. If all right, it will rise a little, and again resume its place.

317. To prove its correctness, just watch the indication when blowing off, and when the steam is down; they nearly all register $\frac{1}{4}$ to 2 lb. in advance, increasing as they get worn out. This can always be counted if it tallies with the blowing-off pressure. It is good practice to have two gauges, one to check the other. If you have not this, borrow one to test your own, but if the rising and falling is fitful and unreliable, have it out and throw it away at once.

Never repair a pressure gauge and trust to it afterwards unless such be done by makers, as the price of a new one is a trifle in comparison with the importance of correctness.

318. It is a very good expedient for protecting them from mud and grit to make a small pellet of brass wire gauze and insert it in the pipe between the boiler and the syphon. This traps gritty matter and prevents choking up of the cavity. This trap should be cleaned from time to time.

Vacuum Gauge.

319. The description of this is the same as for the pressure gauge, with which indeed it is identical; the only difference being that for pressure the instrument is pushed or pressed, whereas for vacuum it is drawn or sucked; the former is above a zero point, the latter under it. For example, a man wheeling a barrow before him corresponds to pressure, and one drawing it after him corresponds to vacuum. Pressure-gauges, so-called, have a stop-pin at zero; to make this into a vacuum or a compound gauge we have only to remove this pin to free the pointer. All gauges for vacuum are marked up to 30 inches, which is the greatest height mercury will rise in the barometer. Every inch of barometer is equal to $\frac{1}{2}$ lb. pressure, therefore, when a gauge shows 28 in., the vacuum will be 14 lb. to the square inch.

Pressure Tests.

320. The working pressure of a boiler should be one-sixth the bursting pressure.

The testing pressure is alterable under certain conditions, but ranges from $1\frac{1}{2}$ to $2\frac{1}{2}$ times the ordinary blowing-off pressure, double the blowing-off pressure being usually adopted.

Bursting pressure of a boiler 18 in. diameter					..	1·416 lb.
,,	,,	,, ,,	36 ,,	,,	..	708 ,,
,,	,,	,, ,,	72 ,,	,,	..	354 ,,

—the thickness of shell in each case being a quarter of an inch. With the thickness of the shell increased to half an inch, the pressure would be exactly double. With the thickness of the shell decreased to one-eighth, the strength would be exactly one-half of the above.

Boiler testing is safest by hydraulic pressure.

Boiler testing by steam pressure—" diamond cut diamond "—is like committing upon it wilful injury. A satisfactory test means a severe one, and a severe test is an ominous one. Putting an unusual strain on a boiler means permanently straining it.

When the test pressure is on, the boiler should be tight, and also maintain its shape and form without bulging out, bending, or straining.

The strongest part of a boiler is the solid plate; the weakest the lines of rivet holes or seams. All strength calculations are based on the bursting pressure of these seams.

321. EXPLOSIONS

generally result from errors of construction and staying. Improper setting, mismanagement, misadventure, neglect, and ignorance are prime causes of premature decay, which will account for approximately 90 per cent. of the whole.

322. *Safety valves setting fast* have been often blamed for causing explosions. If such be the case, then the term safety valve is a misnomer, while the question still remains, Can ordinary boiler safety valves not be depended on for safety? If not, we need go no further for causes of explosions.

Safety valves can only set fast in their seats by corrosion, by being welded by contact, or by getting accidentally or wilfully jammed by interposition of muck or grit. Corrosion can only happen from leakage, which very often arises. But how corrosion should be allowed to accumulate to this extent, which could not be done in a day, can only be accounted for by the grossest neglect. Therefore it must be put down as the valve being stuck fast by corrosion, brought on by leakage resulting from neglect.

323. *Valves* cannot be *jammed*, or injured so as to produce jamming, by the pressure of steam alone or by fair means. The common conical fluked valve will, indeed, stand any amount of hard usage. The spindled kinds also can only be disarranged by foul play. It often happens that grit and muck get caught between the valve and seat while blowing off, and this remaining causes copious blowing off, which

can be stopped by taking the strain a little off the valve and turning it round. But the way adopted in many such cases is *force*, which will give a "set" to the spindle; but any such "set" sufficient to paralyse the action of the valve, will leave it permanently blowing off. The cause of jamming in this case is injury to the valve, the result of ignorance; and the same brain that directed such tampering at first would be very likely to catch the idea of "shoring" or wedging it down. Ignorance multiplied by ignorance is often comical, but in this case is serious and culpable. The criminal part of it is not easily brought home, and not easily punished, lying, as it does, oftener with the owner than the attendant. Hanging weights on to a valve, although not always done in ignorance, has not the criminal complexion of wedging or shoring down, which should be severely punished.

324. The idea of a valve getting united, or *welded to its seat*, is inadmissible. An ordinary mitre valve, with a bearing face of about $\frac{1}{16}$ and 45° angle, cannot unite with the seat while there is a single bucketful of water in the boiler; and supposing it were possible for a boiler to be thus sealed, it certainly could not withstand anything like a bursting pressure. It could not withstand even a testing pressure. If we put the thickness of the boiler shell at $\frac{1}{2}$, and the brass faces in contact at mitre as $\frac{1}{16}$, and allowing the partial union of the brass at mitre as equal to half the cohesive strength of solid brass, and putting the tenacity of brass as three-fourths that of malle-

able iron, then the entire resistance of the safety valve, although really united with the seat, would only be about $\frac{1}{40}$ that of the plate, so that it is a safety valve even still.

325. *Over-pressure*, although often quoted, is inadmissible as a responsible cause of explosion, because it cannot be reconciled with "margin of safety," as it looks incredible that boilers working with a safety margin of one-third to six times should burst at all. If a boiler of, say, half plate, and tested to 90 lb., burst at 85 lb., this is not over-pressure.

If this same boiler was grooved one-quarter deep, and burst at 45 lb., the cause was "grooving."

If a boiler were incrusted so that the plates got hot and an explosion occurred, this condition of things would be the cause.

If a boiler burst while blowing off steam full bore (which means that the boiler was generating more steam than could escape), then the cause is insufficient area of safety valve.

If a boiler was tested to 90 lb. and found all right, and next day exploded at 100 lb., that would not prove over-pressure, because there is a recognised margin of safety over and above the testing pressure. Therefore another cause must be looked for.

326. Over-pressure can only take place with, say, the safety valves wedged or shored down, or where a boiler has no safety valve at all, as it might be with a boiler built of the best material and with the best workmanship up to the highest class, and so perfectly in proportion each part to another that no part was a

weak point, this on bursting would be riven at every seam through the lines of holes, because the strength of a seam, so long as holes are put in the plate, must be less than that of the solid plate. But if a welded or otherwise unriveted boiler be similarly treated, the explosion would be terrific, the whole being shattered to atoms. In ordinary circumstances, then, the term "over-pressure" is indefinite as regards the real cause.

327. The causes of premature decay in steam boilers are just as varied and often as complicated as in the human subject, and between the two there is much in common. With the human subject consumption stands boldly out as a cause of premature decay; with the steam boiler "corrosion." Although iron generally is subject to this disease, boilers are so in particular, especially the outside, where it is induced by the nature of the situation and circumstances taking advantage of the susceptibility of the iron itself. Although to all appearance a steam boiler might be indestructible, and the last to be affected by disease, the case is just the reverse, as from the moment fires are lit the effects of unequal temperature, and consequent unequal expansion and contraction of its parts alternating throughout, begin to consume both structure and material (§§ 312, 212).

328. Another danger lies latent in the contents themselves, which, although a body at rest, are liable under certain circumstances to be provoked into revolt, and to become a body in motion. These circumstances—be it suddenness of opening or shutting

valves, or doors, as at §§ 120-2, 247, or sudden motion of any kind, find response as a matter of course in the body of the contents. This response, being steadily accelerated, becomes a multiplication of energy as a pent-up force, and if the exciting cause be not withdrawn will burst out in priming, and, under sufficiently aggravated circumstances, in rupture or explosion; just as the otherwise docile elephant will run amuck (if provoked) until killed or otherwise disposed of. A gale of wind playing upon the smooth surface of the ocean very soon produces an ocean storm, identical with pressure in a boiler. And the successive states of the ocean—at one time calm, and at another in storm; and of the elephant, once docile, now infuriated—are each typical of the stages of the boiler contents; static pressure at first—after development, a dynamic force.

Any motion, such as pumping or measured shocks, such as dancing to a tune, excite. But above all the blow or measured blows from a hammer (one of which might act as a spark to explode the whole), are incentives to explosion, priming, or overstraining.

We may prove the truth of these remarks by a reversal of the conditions—the reduction of dynamic force to static pressure. By arresting the momentum of a railway train, either against a fixed object or another train, we would have an effective example of the effect of suddenness in arresting force, and when this is joined to suddenness of expansion or contraction, such as opening steam on to frozen pipes, or emptying the steam out of one boiler into another which is cold, such sudden changes of temperature form often all the difference between static pressure and dynamic force.

329. As already shown, the bursting of a boiler is a question of strength as compared with the maximum pressure, and a boiler weakened at any point by decay to half the thickness of the plate will only carry with safety half of the pressure. As the solid plate is from 25 per cent. stronger than the seams, the plate might be decayed to this extent without impairing the pressure strength of the boiler. Explosions, properly so-called, can only take place when the weakness is general, as in lines of seams or weak parts, and is preceded by rupture. Then the escaping steam or water gets a face outside of the boiler to work from as a fulcrum, just as Archimedes would have lifted the world. The effects of an explosion on plates and other parts is not only a question of steam pressure, but of how that force may have been accelerated at the part or multiplied by leverage.

We have explosions from time to time and often of great violence, where none of the causes here assigned were present. We have the experience of a whole series of six good boilers working side by side bursting in rapid succession, spreading death and destruction around. There have also been cases of boilers giving way under pressure little above half of what they had been tested to shortly before.

Altogether we have a heavy record of explosions from unexplained causes.

330. Scientists and learned writers on the subject repudiate the possibility of explosions from any obscure cause not by themselves understood, excepting such stubborn facts as the Liedenfrost phenomena, &c.; and they state the whole subject as one of two sides, positive and negative, viz., the scientific or understood side, which is true, and the superstitious or ignorant side, which is not true. This is simply to say, "All research into this great subject must now close, the science itself being finished"; to ignore the yet undefined causes of general and local eruptions and storms that come and go, whence and whither we know not. The reproduction of these assertions by other writers resemble echoes of a false call; producing a mixture of repudiation, admission and assertion.

331. The Liedenfrost phenomenon is not the only unexplained cause of explosion. It is only just to say that unsupported repudiation is as unreasonable as unsupported assertion, and that what is comprehensible ought to be at least explainable. Explosions are in no way affected by popular fancy or fashion. And, just as sure as a few degrees of temperature, top or bottom, exceeding previous experiences, will raise new issues in many branches of science, so will the continual changes of ideas regarding construction, pressure, temperature, and kindred facts innovate on the science relating to steam boilers. There will always be something left for the rising generation to investigate and

add to history. The belief in causes of explosion not yet defined is put by many on the same shelf with the ancient belief in ghosts as being out of fashion, like other vulgar things that cannot be reconciled with the intellectuality of the times. But there can be no improvement in rejecting the ancient ghosts and accepting the more fashionable spirits—the one being just as irrational as the other is superstitious.

332. It is perfectly rational to say that the same agency that would burst a boiler would also prime it; just as a force that would prime a boiler would, if sufficiently developed, burst it.

There are often forces in operation inside a boiler which are not bargained for, and it may be safely assumed that dynamic developments, either of themselves or compounded with other forces, will often increase the strain at one part equal to several times that of the pressure. Such intense local strain may come or go like the blow of a hammer, and every one must be acquainted with the effective result of the blow of a hammer as compared with simple pressure, increasing in effect the pressure at one point to several times the working pressure; and pressure, like sound and light, requires time to travel, so that before the safety valve is reached the explosion has occurred.

333. We may also account for that kind of explosions that occur when a boiler is "rushed," to raise steam quickly. Favourable for such a result would be an upright boiler with very little, if any, water over the crown, and a forced fire of sticks piled up-

right, or bags, shavings, grease, and such like. The concentrated effect of the flame under such conditions would be such as to drive the water from the plate. This action and consequent reaction would establish a pulsation of the contents which would be similar to the action of a pump. The effect of this is exactly the same as throwing water on a hot plate. And it has been proved that steam raised from water thrown upon a hot plate will rise from 1 to 12 atmospheres in a minute (Fig. 5).

334. This raising of steam from a hot plate is demonstrated wherever fire and water or a hot plate and water meet; the steam thus produced, being what may be termed super-generated steam, is to common steam in point of sharpness and energy as nitro-glycerine is to gunpowder. Another link of the same chain is the decomposition of water by the loss of its fixed air (§§ 337, 178).

Every blacksmith knows, when he cools his hot tongs in the trough, of the dynamic convulsions produced, feeling to the hand as severe electro-galvanic shocks. Blacksmiths also know, if they strike their hot iron over a drop of water put on the anvil, it will change in an instant from the fluid state into the gaseous with a report similar to that of a cannon.

Every boiler attendant or fireman knows the effect of cold water thrown over hot coals, and the explosive rumbling noise accompanying it.

Every engineer in charge of boilers knows (when steam is let into cold pipes, or cold water into hot pipes) of the violent "crack, crack, cracking," and starting of joints that follows, and often the bursting of the pipes themselves—the result of the antagonism of these forces. When such effects are produced, these forces simply neutralise or counteract each other. What may we expect when the turn comes for their energies to be united?

335. Glancing back, we find our first wonder was how steam boilers with such a liberal margin of safety could burst at all. But now, with the reasonings of such facts and collateral proofs on either side, the wonder becomes how steam boilers can stand at all, or that explosions are not more frequent. The margin of safety appears like a farce. We wonder how the whole subject of steam-boiler manufacture, testing, inspecting, attendants, &c., has not been taken up by the demand of the people, or that this important subject has not yet formed a plank in some political platform.

Under ordinary circumstances, there is nothing unusual to complain of, but when any condition gets to be abnormal, that becomes a joint in the armour, —at once a weak and a vital point. Scientists and learned writers altogether are like a pack of hounds at fault, or an actor who has forgotten his cue. It is not single abnormal conditions alone that claim attention, but when these are compounded, the science itself becomes a series of paradoxes.

336. A few years ago a terrible accident occurred at Corbeil in France. A small steamer was frozen up; three men were sent to raise steam and have her removed. As soon as a little heat was generated, a

fearful explosion took place, supposed to be owing to the steam pipes being filled with ice. All three were terribly injured, and the steamer sank immediately.

337. It has been found that water deprived of its fixed air has one peculiar property among others, viz. that it may be heated several degrees above the normal boiling point without showing signs of ebullition; but when a certain temperature is reached by the continual application of heat, the whole mass suddenly becomes steam, the fixed air of the water being displaced or decomposed by the freezing, termed the spheroidal state of water.

Experiment A.

338. Put a basin of water outside the house on a frosty night and let it freeze. Pour as much oil on it as will cover the surface of the ice; then set the basin where the ice may melt; when all is melted, draw off by means of a syphon, say, half a pint from under the oil; if this is put into a saucepan, having its lid on, and placed on the fire, when you think it ought to begin to boil, stand clear, for there will be a repetition, on a small scale, of the Corbeil explosion. If tried by a thermometer, 216° will be shown before there is any danger of explosion, but it would not be advisable to watch for ebullition after this temperature is reached.

Experiment B.

Mr. Burret, F.E.S., read before the British Association a paper on the Spheroidal State of Water and its relationship to certain Boiler Explosions. He says:—"I plunged a red-hot ball of copper into water where I had washed my hands; it made no hissing or visible evolution of steam. Using plain water, the hissing was loud."

As the ball cools mainly by reaction, the shell of vapour grows thinner, then collapses with a loud report, producing volumes of steam *and broken glass.* The cause of this phenomenon was the soap in the water. The same effects would also be produced by any fatty matter or oils (§ 346).

Experiment C.

Take a very thin vessel and fill it with boiling water, and make the opening steam-tight. If cold water be externally applied, an explosion will immediately occur. If boiling or warm water be used, it produces no effect. Some writers attribute the cause of the vessel exploding or collapsing when immersed in cold water to the sudden cold producing a vacuum —being about 70° F.

339. These experiments all demonstrate that in matters of boiler explosions much has yet to be known, much more light requires to be shed on the subject. It is worthy of remark that all these phenomena are produced by (to us) abnormal conditions of temperature.

We know, for instance, that a few degrees above or below our maximum limits of temperature would bring about such abnormal experiences at every step as to paralyse all written science. The question is, if we could ourselves stand the change and still exist to bear further witness, or if we are not trespassing into higher or lower replicates of nature, which are outside of our own natural precincts, and for which we are neither physically nor intellectually fitted. Indeed, even within our own realm and its sciences, how often we are at fault! We can always understand things that run parallel, and also things that taper to a point, but we cannot explain the workings of wheels within wheels, or the master wheel itself. Also, for example, if we take the well-known science of music, the scale of the human voice, why the note intervals are irregular throughout, no man can tell, and why certain sounds harmonise and others jar, no man can tell. Let us, from a standpoint of blood-heat, put one hand in cold water and the other in hot water, if the degrees each way be uniform, our sensations will be alike. If we could stand at the equator and put a hand on each pole we might thus prove that in this way, extremes meet for ought we can gainsay. The same plan of intervals, octaves, periods, or replicates, seems now, as from all time, simply as nature repeating itself, and still the same principles of development and decay prevail. Therefore it is not incumbent on practical authorities on boiler explosions to have popular causes for such occurrences.

340. The subject of latent heat is a very striking illustration of these remarks, inasmuch as we can descend perhaps through many repetitions of nature ere we arrive at absolute zero, while the same would hold good were we to ascend.

341. Joints and Connections,

especially man-hole and sludge doors, require both skill and care. The best jointing in use is common Tuck's packing, made with a long scarf, and served along the centre, which is better than tying. The joint should be as soft as the other part. Tying them round makes them hard, and when flattened by tightening, the twine must either burst or cut the packing as it spreads. One side should have some jointing of white and red lead to make it adhere where it can best remain. The other side should be rubbed with a mixture of black lead and tallow; the tallow keeps it soft and free, the black lead preserves it from heat, and along with the tallow keeps it from adhering. Treated in this way joints may be used from three to six times, or so long as they keep together, otherwise the heavy expense is a veto upon boiler examination.

342. Blow, scum, and all other cocks should be packed with Tuck's or asbestos packing, with plenty of black lead mixed with a little tallow. For flange joints about a boiler, if the flange is well fitted to the plate, light canvas steeped in stiff lead-paint should be used, or one to two thicknesses of gauze wire with a coating of cement between. When using any kind of oil cement, rub, first of all, the faces to be jointed with a little linseed oil, which facilitates the spreading and working of the putty. About the strongest and most generally applicable jointing putty to be had is the common moist white lead,

stiffened to the consistence desired with dry red lead.

343. Canvas joints steeped previously in thick paint, or faced with thin cement, are well adapted for sea cocks and ship's-side work, where there is not great heat, otherwise gauze is preferable.

344. Asbestos jointing cannot be excelled for a general handy steam jointing; but there must not be water for joints and packing for cocks on boilers above the water-line. Scraps of it can be utilised for packing cocks large or small. Asbestos jointing need only be steeped a short time in water in order to be sufficiently swelled for the purpose. It can be used over and over again by steeping, and as a joint is imperishable. A little black lead peppered over it before closing prevents its adhering, the joint being left intact. When brushed over with water to soften it, it is again ready for use. Any little scrap comes in handy for repairing.

345. The great object in all jointing is to have clean, flat, and well-fitted faces. Have the bolt-holes plentiful, large enough, with room for turning nuts, and what is very important, all joints, gauze, canvas, or asbestos should have their holes also large, because very little doubling of the joint at the bolts will cause a bad job, which happens very often from this cause alone. Putty and lead-wire or twine are not suitable for boilers. Firstly, if the surfaces are not convex they are not to be depended on otherwise as being joinable faces. Secondly, twine does not stand the heat. Insertion is best

suited for copper piping with good faces; it will stand either heat or water. Lamp wick ought never to be used for packing or jointing about a boiler.

346. Cylinder Lubricants

often do more damage to the boilers than good to the cylinders, from the acids they contain. All oils that solidify or decompose at a low temperature with a heavy residue are very bad for both engines and boilers. If heavy, and seeking downwards towards furnace crowns, they deposit there. Mineral oils of a kerosene nature that keep clean cylinders, are apt to make a dirty boiler with the grease, or a rusty boiler from the presence of acids.

It must be borne in mind that all oily or fatty matters or cylinder lubricants whatsoever, are injurious to the boiler, and have to be looked out for there as a spent boiler compound (§ 205), besides the other evils referred to. Therefore, the injecting of tallow and other fatty matter into boilers to allay priming is certainly choosing the greater of two evils, as, although the tallow would actually stop the priming the matter does not end there (§ 207), but a course of mischief would be begun. All such substances contain acids which are more or less causes of pitting, afterwards finding their way upwards to the surface as *scum*, which although, chemically speaking, it perish, will during priming incorporate with the mud from the bottom, assisting it to float and deposit, not to mention the mechanical evil *mud trap* (§ 231). Ejecting this by the scum cock is also a serious loss of fresh water.

PRESSURES.

347. Respecting pressures much misconception has always prevailed, especially with young engineers. The terms, "gross pressure," "total pressure," "absolute pressure," &c., generally all signify the same thing. It is also authoritatively laid down that there is no such thing as *suction*. Also that neither steam nor vacuum gauges indicate correct pressure, or vacuum, and that to find out steam pressure we must have recourse to the barometer.

Now, however literally correct all this may be, it is vague and practically misleading, and, indeed, not one engineer amongst a thousand one may meet could tell what the terms gross and absolute pressure meant. It is much the same now as in the atmospheric engine times of old, with a barometer as the No. 1 piece of engine-room furniture. Although the atmospheric engine is now obsolete, and the barometer is no longer known in the engine-room, or understood by the average engineer, the language is still extant, but still it is not read. Rough-and-ready practical rules can only be tolerated. Nothing is so "bamboozling" and disgusting to a young engineer after a breeding of conventional discipline and technical hair splitting, as to find his teaching not conformable with practice.

Therefore, to make the matter acceptable at all, we must assume certain reasonable conditions, viz. that pressure as indicated by the pressure gauge is really correct, and that vacuum as indicated by the vacuum gauge is also correct, and that there is suction.

348. Our faith in the indications of the pressure gauge is owing to its being specially made for the purpose, that it is applied directly to the contents of the boiler, and that it has a record of many years' faithful service, and will indicate all pressures, whether steam, air, or water if pressure exists, and will also indicate the presence of vacuum as faithfully as pressure. The instrument is the same for one or other, only the application is inverse. Nothing makes a subject more intelligible than the apparent absence of ambiguity.

A B is atmospheric or zero line, a fixed point; O V is the limit of vacuum, variable with the weather; $b\,p$ representative of back pressure; P P back pressure. All above the line A B is pressure; all below the line A B is vacuum. $b\,p$ means back pressure, which is a mixture of pressure and vacuum. The pressure line S indicates simple pressure, the same as the pressure gauge and simple vacuum. The pressure line T indicates the same with the pressure of the atmosphere added, as from line A B to zero. The space from B to zero, or O V, 30 inches, is this: 30 inches being the maximum height, the mercury will rise from the pressure of atmosphere. As 2 inches of mercury rise equals 1 lb. of pressure, we then have 15 lb. pressure of atmosphere as marked on line A B S. Sometimes the mercury only rises 28 inches, which equals only 14 lb. pressure; 29 inches which equals only 14½ lb. pressure. This

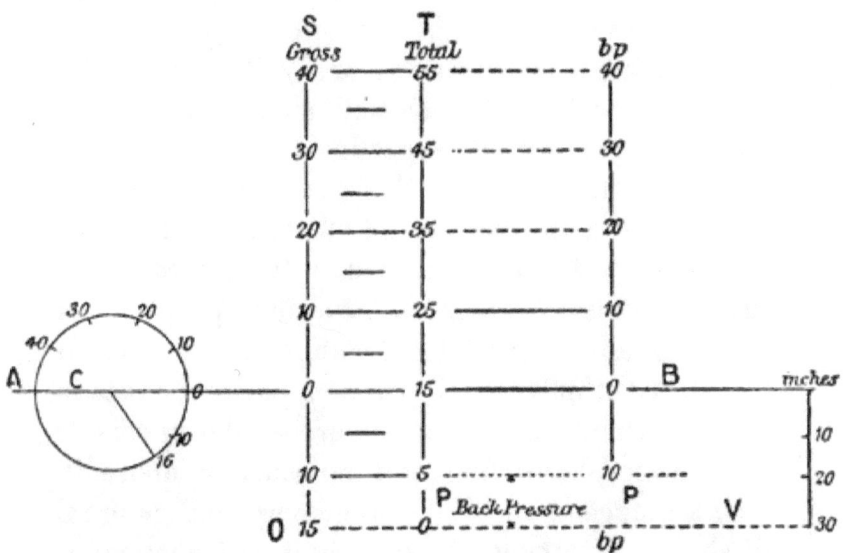

Fig. 34.

barometric or atmospheric pressure is continually varying, consequently this 15 lb. at line T will very seldom be correct, only when the atmosphere is 30 inches, then the measurement will be from vacuum limit line (§§ 350, 222-3).

On the other hand, the pressure line S starts from the atmospheric line as a zero or fixed point, and from this point up indicates real pressure only, no vacuum being added.

349. We must look upon a vacuum as not being an institution in nature, but by nature opposed as a hybrid, and by it hunted down at every point, as if

by a great natural enemy, and always with success. As is well known, to protect or maintain a vacuum perfect is, in consequence of this antagonism, impossible. This success referred to is shown at $b\ p$, and is termed back pressure, as indicative of the amount the vacuum has been diluted ($b\ p$ 5 lb.)

If a vacuum be 26 inches and 7 lb. per inch of steam escapes into it, then this steam will simply dilute the vacuum to 6 lb. per inch.

Again, we must assume this back pressure to be a misnomer, because this 7 lb. of steam pressure, when it escapes into the vacuum, is simply lost, as it now ceases to exist. There is no back pressure or pressure of any kind—all is vacuum; and no matter how diluted it may be, as long as a particle exists in the vessel there can be no pressure on the underside of the atmospheric line, and no vacuum above it, which a glance at the dial of the compound gauge C will show. If pressure rises 1 lb. it will show, if it falls 2 lb. from that there will be 1 lb. vacuum. From our working level of vacuum, say, 26 inches or 13 lb. to zero will always represent how much it has been diluted, either with steam, air, or gas, or, as has been observed, the extent to which nature as an enemy to the vacuum prevails against it, which is the meaning of the phrase *back pressure*.

A vacuum, therefore, naturally has no real existence, but is created by artificial means, viz.: by filling a vessel with steam, thereby expelling the air. On the steam being condensed to a small drop of water (Fig. 16), the space is thus void or a vacuum.

This vacuum-making is one great feature of economy in the use of steam, that it fills a space and voids it, which act moves machinery.

Suppose an empty boiler with the man-hole door off. There is 15 lb. pressure said to be inside and outside of that boiler. The door is put on, and the boiler is filled with steam, which is then condensed, leaving a vacuum inside. What about the pressure then?

The force with which the outside air rushes to fill this void is equal to 15 lb. per square inch. This 15 lb. of advantage is dug out at the expense of 16 lb. of steam, and did not previously exist. With open vessels there can be outside pressure—all being equality. The force in this case was produced by creating a void inside, which is equal to but not synonymous with, a pressure outside. The very same result is obtained by the inverse method, viz.: by creating a void outside the boiler, which is equal to but not synonymous with a pressure inside.

350. Again, when the day's work is over and as the fires die out and steam falls, so will the pointer of the pressure gauge fall to zero. No steam is in the boiler. Open a gauge cock—there is nothing either inhaled or admitted. A few minutes more and air begins to rush by this open cock. The steam is condensing inside of the boiler, which is also shown by the gauge. In the days of atmospheric engines a valve opening inwards was fitted to boilers, called atmospheric valve, which at this point opened by the void inside and equalised the pressure. It is

very bad for the boilers, even of the present time, to allow a vacuum to form inside, it being a cause of leakage (§ 267).

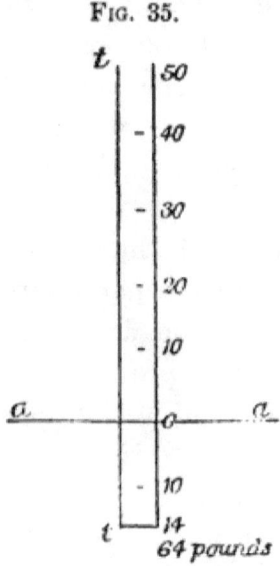

Fig. 35.

Suppose *t t* was a tube, *a a* atmospheric line or zero. This tube is marked off for pressure above the zero line, and similarly below it for vacuum.

> The pressure 50 lb.
> The vacuum 14 lb.
> —————
> Total force   64 lb.

Five lb. escapes from the steam to the vacuum side, which is 5 lb. less steam and also 5 lb. more vacuum, equal to 10 lb. less force, the same as would result from a weight lifted out of one scale and placed in the other.

INDEX.

The figures refer to the numbered paragraphs.

A.

Abnormal conditions, 335
Absolute zero, 214
Accumulants, 85, 181, 208
Acidity, 178–184
—— causes, 181–183
—— disposition, 179
—— germ, 181
—— peculiarities, 182
—— proofs, 179–182
—— remedies, 183–184
Acids, 44, 181
Agitation of water, 265, 266
Air and coal gases, 19–21
—— composition, 21
—— expansion, 19
Alabaster, 45
Alarums, 259,
Alkalis, 44
Alum, 45
Alumina, 45
Aluminum, 45
Amalgam, 201
Ammonia, 55
Analysis, Thames water, 43, C
—— boiler scale, 57
Analysis, sea water, 61, 43, AB
Ant's burden, 41, A
Artificial scaling, 102–105
Asbestos, 344
Ashes, 139, 239, 241
Ash-pits, 241
Astringents, 75
Atmospheric valve, 350
Attendants, 225, 226, 335
Auctioneer's licence, 218
Automatic firing, 33

B.

Back smoke, 248
Baffled deposit, 166
Baking scale, 166
Bane and antidote, 190, 204
Banked fires, 274, 275
Barnacles, 173
Barometer, 69, 347, 348
Bars coaling, 235
—— dropping, 250–252
—— —— cause, 252
—— —— prevention, 251
Berthier's table, 61

Index.

Blacksmith's outlay, 182
Blowing, 72–76, 161–177
—— cautions, 175
—— directions, 165–175
—— down, 165–167
—— down for scaling, 166
—— out water, 78, 101, 161, 167
—— practices, 72–76
—— queries, 80–81
—— schools, 72–77, 162
Blowing-off cocks, 161, 342
—— -off pressure, 142
—— -off steam, 261
Boiler attendants, 225, 226 335,
—— barnacles, 133
—— cleaning, 173, 174
—— construction, 96
—— Cornish, 301
—— decay, 327
—— digestion, 131
—— dirty, 173
—— donkey, 132, 307
—— drying, 166
—— egg-end, 300
—— examinations, 168
—— freshening, 101, 161
—— filling, 164, 228
—— forcing, 141
—— functions, 131
—— gaining water, 262, 263
—— humours, 131
—— individuality, 226
—— larger, 144, 244
—— leakage, 264
Boilers lifting water, 232
—— locomotive, 204
—— marine, 89, 136
—— medicines, 207, 208
—— new, 89
—— nondescript, 226

Boilers, plates burning, 208
—— —— buckling, 208
—— running out, 166
—— "rushed," 226, 333
—— rusting, 270
—— scale, *see* scale
—— sickness, 114
—— sounding, 301
—— upright, 298, 299, 307
—— vacuum, 267, 268
—— washing out, 173
Boiling points, 69
Bottom mud, 115
Boy philosophers, 213, 214
Bridges, 286, 289
—— too high, 289
—— too low, 289
Buckling plates, 276
Building and knocking down, 27, B
Bulk, 221, 222
Bunker space, 144
Bunsen burner, 249
Burning building, 23, A
—— cinders, 242
—— down fires, 236
—— plates, 208, 276
Bursting pipes, 334
—— tubes, 276–279, 280

C

Caking coal, 15
Calcium, 45
—— chloride, 48
Candle flame, 25, B
Carbon, 46
Carbonic acid gas, 22, 30, 51, 52
—— oxide gas, 22, 30
Carbonate, 48, 55, 51

Index.   223

Carbonate lime, 47, 51, 58, 61
—— magnesia, 47, 51, 59, 61
—— soda, 48
—— zinc, 198
Caustic soda, 195
Changing engineers, 139
—— water, 120, 165, 183
Chalk, 45
Characteristics of air, 19, 27
—— coal, 13, 28
—— steam, 223
—— water, 8
Charring coal, 25
Checking fires, 256
Chemical action, 200–202
—— combinations, 47–52
—— combining quantities, 54, 55
—— compounds, 47, 205–208
—— exchanges, 187, 190, 197, 204
—— virtues, 186
Chimneys, clean, 149
—— cowls, 245
—— dirty, 149
—— top ablaze, 249
Chipping heating surfaces, 271
Chlorides, 55
Chloride sodium, 47–49, 60, 61
—— magnesia, 47, 50, 61
—— calcium, 47
Chlorine, 46
Circulation, 37, 96, 127, 130–132
—— inherent, 38
—— sluggish, 129
—— throttled, 128
Cinders, 242
City smoke, 24
Cleaning fires, 139, 237

Cleanliness, 173
Clinker, 239
Close furnace, 26
Coal, 13
—— analysis, 19
—— on bars, 235
—— bed burnt, 13
—— bituminous, 17
—— cleavage, 16
—— caking, 15
—— coking, 13
—— composition, 19,
—— deterioration, 16
—— gases, 19, 20, 21
—— kinds, 13, 14
—— patent, 18
—— preservation, 16
—— slack, 16, 18
—— smokeless, 13
—— sources of loss, 21
—— vapour, 27
—— watering, 13, 14
Coasting steamers, 176
Cold air rush, 27, 242
Collapse, 92, 309
Combinations of coal and air, 28
—— of solids, 61–66
Combining quantities, 30
Combustion, 22–33
Common lime deposit, 80–92
—— salt deposit, 92–95, 306
Compounds, 42, 205–208
Concentrated flame, 285–297, 333
Construction, 96
Consumption and speed, 150–156
Corbeil explosion, 336
Cornish boilers, 212, 301

Corrosion, 269, 270
Corrugated furnaces, 297
Cylinder lubricants, 346

D.

Dampers, 253–257
Danger of neglect, 128, 159, 160
Dead reckoning, 97
—— weight, 261
Decomposition of water, 61, D, 334
—— zinc, 201
Density, 79
Depositions of solids, 80–100
Deterioration of coal, 16
—— zinc, 201
Dirty water, 117
Distilled water, 178
Doctor and patient, 126
Donkey boiler, 132, 298, 299, 307–312
—— man (*see* boiler attendants)
Down-draught, 246, 247
—— abatement, 247
Draught, 96, 244, 293–296
—— creation, 20, 21, 30
Dregs, 181
Dropping fire bars, 250–252
Dynamic development, 332–334
—— force, 120, 121, 247, 328
—— convulsions, 334

E.

Easing fires, 256
Economy, 144
Effect of density, 92–95

Effect of dirty water, 117
—— latent heat, 225, 226
—— scale, 150–157
—— shortness of water, 92–5
—— soda, 193
—— soot, 150–157
—— rolling ship, 261
—— zinc, 193
Egg-boiling, 118
Egg-end boilers, 133, 300
Ejecting mud and water, 180
Electrogenes, 200
Elements of nature, 44
Engineers, nondescript, 76
Engineers, changing, 139
Escaping forces, 291, 292
Evaporation of coal, 16, 30
Exchanges, 187, &c.
Expansion of air gases, 19
—— of boilers, 92, 212
—— steam, 222, 223
—— water, 164, 209–211
Explosions, 321–340
—— supergenerated steam, 226

F.

Fair exchange, 42
—— steaming, 154, 155
False bottom, 266
—— water, 230
Faraday's tables, 67
Feed-water, 158–160
—— —— treatment, 41
—— —— not returned, 91
—— —— loosing, 264
Filigree work, 179, 182
Filling boilers, 228
Fire bars dropping, 250–252

Index. 225

Fire and water contest, 92
Fires, burning down, 236
—— cleaning, 236, 237
—— common, 26
—— working, 235-238
Fireman, 225, 237
Firing, 234-243
—— automatic, 33
—— *versus* combustion, 240
Fixed air, 66, 178
Flame concentration, 264
—— spreading, 287-289
Flanged joints, 342
Flogging tubes and flues, 168
Fluxed surfaces, 104
Focal distance, 293-296
Forcing boilers, 141, 226, 333
Frequent stoppage, 89, 257
Freshening the water, 77, 101, 161
Fresh water at surface, 35, 175, 299
Frost results, 211, 215
Funnel gases, 20, 249
—— top ablaze, 249
Furnaces, 26, 29, 233
—— collapsing, 92-96, 310, 311
—— corrugated, 297
—— large, 242
Fusible plugs, 259

G.

Galloway tubes, 290-299
———— flogging, 168
Galvanic action, 158
Gas combinations, 27, 28
Gauge glass, 229
—— pressure, 221, 347, 348

Generation of steam, 34, 35
Green fire, 29
Grooving, 313
Gross pressure, 220, 221

H.

Handy steam, 88, 258
Hard firing, 139
Harmless deposits, 192
Heat, 44, 219-226
—— interception, 297
—— latent in water, 216, 217
—— —— in steam, 217, 218
—— non-conduction, 145-147
Heating surfaces, 271-273
Holing fires, 275
Hot ball experiment, 338
Hot plate, 37
Hydrate, 55
Hydrochloric acid, 47-49, 285
Hydrogen, 14, 46
Hydrokinator, 265, 266

I.

Ice density, 67-101
—— properties, 215
Ignition point, 25, 29, B
Ignorant treatment, 71, 76
Impregnated waters, 10-12
Inattention, 128, 159, 160
Increasing consumption, 154-156
—— speed, 141
Incrustation, 36-42
Inertia, 154, 155
Ingredients of air, 22

Q

Ingredients of coal, 19
—— scale, 56, 57
—— scum, 346
—— water, 43, 61
Inherent circulation, 38, 132
Institution in nature, 109, 349
Insulation, 157
Intake pipe, 122
Interception of heat, 297
Interposition of soda, 186
—— of zinc, 197
Iron plate attacked, 179, 199, 285
—— —— rolled skin virtue, 179, 314

J.

Joints and connections, 341–346
Joule's equivalent of heat, 213

L.

Land boilers, 300, 301
Large   ,,   144
Latent heat, 213–226
Leaky boilers, 264
—— condenser, 262, 263
—— tubes, 276–284
—— pipes, 334
—— tubes stopping, 281–284
Liedenfrost phenomena, 330
Lime carbonate, 47, 70
—— extraction, 41
—— nitrate, 47, 85
—— sulphate, 56, 70
Locomotive boilers, 121, 122, 204
Lubricants, 346

M.

Magnesia, 45
—— sulphate, 61
—— chloride, 50
Magnesium, 45
Mal-construction, 127
Manipulation of flame, 287, 288
Marble, 45
Marcet's tables, 68
Margin of safety, 335
Marine boilers, 136
Mean pressure, 142
Mechanical virtues, 96
Medicines, 207
Metals, 45
—— non, 46
Moss or bog water, 133
Mud ejecting, 115
—— in bottom, 230
—— line, 230
—— trap, 231
Multitubular boilers, 136
Murdered fuel, 139, 152–155
Muriatic acid, 50

N.

Neglect, 128, 159, 160
Nitrates, 55
—— of lime, 47, 85
Nitrogen, 46
Non-conduction, 145–148
Non-metals, 46

O.

Oak bark, 135, 136
—— sawdust, 133–136
Ocean travellers, 89, 137

Over-pressure, 325, 326
Over-thickness of plates, 285
Oxide, iron, 47, 270
Oxygen, 46, 270

P.

Paraffin lamp, 29
Parasites, 133
Patent fuel, 18
—— stoppers, 281-283
Peculiarities of boilers, 103
Pigeon holes, 247, 251
Pitting, 133, 134
—— prevention, 136, A B, 174
Phosphates, 55
Phosphorus, 46
Plates, buckling, 208, 276
Plurality of boilers, 159, 160
Potash, 45
Potassium, 45
Pressure, 317-350
—— gauge, 316-318, 347-350
—— gross, 225
—— relations, 111, 348, 349
—— safe, 320
—— static, 120-124, 224
—— tests, 180, 320, 332
—— unequal, 127, 180, 332
Priming, 109-126, 164, 180

Q.

Quacks, 71

R.

Rain-water, 204
Raising steam, 209

Ratio of friction, 154, 155
—— of progress, 27, B
—— of solids, 61-66
Reaction, 42
Rignault's tables, 225
Rejecting mud, 115
Repairs, 302, 303, &c.
Repetitions and reproductions of nature, 109, 339
Resuming of solids, 87, 164
Retention of water, 176, &c.
Running-out boilers, 165, 166
Rupture, 93, 328
Rushing boilers, 141, 226, 333
Rust, 105, 108, 133, 269, 322

S.

Safe pressures, 320
Safety margin, 335
—— valves, 260, 261, 322-324
Salinometer, 76, 80, 97
—— *versus* acidity, 182
Salt, common, 81
—— deposit, 91, 100, 306
—— fluxed surfaces, 92
—— seasoned, 105-108
Salted job, 139
Saltness, natural, 81
Saltpetre, 47
Salts, 44
Saturated steam, 158
—— water, 61
Scale amalgam, 104
—— analysis, 56
—— artificial, 102-105
—— baking, 89
—— character, 85, 188
—— chipping, 271-273

Index.

Scale compositions, 71, 205
—— contracting, 37
—— deposit, 1st lime, 80–92
—— —— 2nd salt, 92, 306
—— expansion, 169, 170
—— hardening, 87, 166
—— leakage, 276
—— prevention, 71, 204–208
—— promotion, 37, 88
—— removal, 271–273
—— *versus* soda, 188, 189
—— softening, 166
—— sovereign cure, 204–206
—— special, 286
—— soot, 150–157
Screenings, 32
Scum, 161, 231
Scumming, 161–177
Sea water, 43, 61, 67, 210
Sediment, 10–12, 209–211
Sensations of priming, 100
Sensible heat, 214–226
See-saw, 213
Silica and silicon, 46, 47, 55
Slack coal, 32
Sludge doors, 341
Small coal, 32
Smoke, 29, 243
Soap bubbles, 34
Soapy water, 134
Soda crystals, 133, B, 185–195
—— carbonate, 48
—— sulphate, 47, 49
Sodium, 45
Sodium chloride, 48
Soft water, 9
Solids, 71, 87, 164
Solutions, 209, 210
Soot, 145–157
—— patches, 119

Sounding boilers, 301
Sour water, 178
Special scale, 286
Specific gravity, 67–69, 79–81, 209
Speed and consumption, 141, 152–157
Static pressure, 328, 332
Steady steam, 235
Steam, 34
—— expansion, 222
—— gauge, 316–318
—— handy, 88, 258
—— latent heat of, 213
—— raising, 209
—— saturated, 158
—— space, 158
—— temperature, 80–83
Steamers, coasting, 176
—— river, 89, 161, 176
—— ocean, 177
Stiff for steam, 137–144
Stirring the pan, 36
Stoppage *versus* inertia, 154, 155
Stoppages frequent, 89, 257
Storm path, 112
Stove, common, 149, 157
Substances in nature, 44
Supergenerated steam, 37, 226
Super-heating, 227
Sulphate, 55
—— copper, 47
—— lime, 47, 57, 70
—— magnesia, 61–65
—— potash, 47
—— soda, 49, 61–65
Sulphide of lead, 47
Sulphur, 46
Sulphuric acid, 49, 285

Index. 229

Sulphurous coal, 53
Supplementary feed, 159
Suspension, 209, 210

T.

Tallow, 126, 346
Tally keeping, 97–99
Tannic acid, 133–136
Tear and wear, 212, 312
Temperature, 214
—— steam, 221–223
Tendency to repel scale, 89
Testing boilers, 320
—— salinometer, 98
—— water, 9–11
Thermal units, 213
Thermometer, 69, 220, 221, 222
Thief to catch thief, 41
Trial trips, 137
Tubes, 276–284, 289
—— pitting, 135, 136
—— scale, 145–156
—— scaling, 90–93
—— spaces throttled, 127–129, 150

U.

Unequal draught, 245, A
—— expansion, 212
—— patches, 118, 119
—— pressures, 127–129, 332
—— temperatures, 103
Ulcers, 133, 135
Upright boilers, 298, 306

V.

Vacuum, filling boilers by, 267–268
—— gauge, 319
Values of heating surfaces, 273
Valves, atmospheric, 350
—— safety, *see* 'Safety Valves'
Ventilation, 245–248
Virtues of construction, 96
—— soda, 185
—— zinc, 196

W.

Wash-basin, 134
Washing out boilers, 173
Water, 8–12
—— acids, 44, 181
—— acidulated effects, 178
—— analysis, 43, A B, 43, C, 61
—— and fire contest, 92
—— boiling points, 69, A
—— changing, 120, 165–168, 183
—— characteristics, 8, 210, 211
—— composition, 43, 61, 210
—— density, 209–211
—— dirty, 10
—— distilled, 178
—— expansion, 209–211
—— falls, 291, 292
—— false, 230
—— impregnated, 209
—— *versus* frost, 211
—— without fixed air, 178
—— level, 229
—— medication, 41

Water mossy or bog, 10, 133, 204
—— rain, 11, 204, 210
—— saturation, 210
—— sea, 43, 61–67, 210
—— soft, 11, 204
—— solids, 10–12, 164, 209–211
—— sour, 178
—— tests, 9–11
—— Thames, 43

Water unfitness, 12
Wear and tear, 312
Watering coal, 13, 14

Z.

Zinc, 196–203

LONDON:
PRINTED BY WILLIAM CLOWES AND SONS, LIMITED, STAMFORD STREET
AND CHARING CROSS.

www.ingramcontent.com/pod-product-compliance
Lightning Source LLC
Chambersburg PA
CBHW021802230426
43669CB00008B/601